A DECADE OF PROGRESS AFTER THE FUKUSHIMA DAIICHI NPP ACCIDENT

PROCEEDINGS SERIES

A DECADE OF PROGRESS AFTER THE FUKUSHIMA DAIICHI NPP ACCIDENT

BUILDING ON LESSONS LEARNED TO FURTHER STRENGTHEN NUCLEAR SAFETY

PROCEEDINGS OF AN INTERNATIONAL CONFERENCE
ORGANIZED BY THE
INTERNATIONAL ATOMIC ENERGY AGENCY
IN COOPERATION WITH THE
FOOD AND AGRICULTURE ORGANIZATION OF THE UNITED NATIONS,
INTERNATIONAL LABOUR ORGANIZATION,
ORGANISATION FOR ECONOMIC CO-OPERATION
AND DEVELOPMENT/NUCLEAR ENERGY AGENCY,
PREPARATORY COMMISSION FOR THE COMPREHENSIVE
NUCLEAR-TEST-BAN TREATY ORGANIZATION,
UNITED NATIONS SCIENTIFIC COMMITTEE ON THE EFFECTS OF ATOMIC RADIATION,
WORLD HEALTH ORGANIZATION
AND THE WORLD METEOROLOGICAL ORGANIZATION
AND HELD IN VIENNA, AUSTRIA, 8–12 NOVEMBER 2021

INTERNATIONAL ATOMIC ENERGY AGENCY
VIENNA, 2023

COPYRIGHT NOTICE

© IAEA, 2023

Printed by the IAEA in Austria
July 2023
STI/PUB/2061

IAEA Library Cataloguing in Publication Data

Names: International Atomic Energy Agency.
Title: A decade of progress after the Fukushima Daiichi NPP accident / International Atomic Energy Agency.
Description: Vienna : International Atomic Energy Agency, 2023. | Series: Proceedings series (International Atomic Energy Agency), ISSN 0074–1884 | Includes bibliographical references.
Identifiers: IAEAL 23-01597 | ISBN 978–92–0–132223–4 (paperback : alk. paper) | ISBN 978–92–0–132123–7 (pdf)
Subjects: LCSH: Fukushima Nuclear Disaster, Japan, 2011. | Fukushima Nuclear Disaster, Japan, 2011 — Conference proceedings. | Nuclear power plants — Japan — Safety measures. | Nuclear accidents — Japan. | Nuclear industry — Safety measures.
Classification: UDC 621.039.588 (520) | STI/PUB/2061

FOREWORD

The International Conference on a Decade of Progress After Fukushima Daiichi: Building on the Lessons Learned to Further Strengthen Nuclear Safety was organized by the IAEA and held in Vienna in November 2021. The conference reviewed lessons from the accident at the Fukushima Daiichi nuclear power plant and the actions taken worldwide in the 10 years after the accident to strengthen nuclear safety. These actions included the development of the IAEA Action Plan on Nuclear Safety, endorsed by IAEA Member States in 2011, which defined a programme of work to strengthen the global nuclear safety framework in response to the accident; the publication of The Fukushima Daiichi Accident — Report by the Director General and its five accompanying technical volumes in 2015; the adoption of the Vienna Declaration on Nuclear Safety in 2015 in accordance with the objectives of the Convention on Nuclear Safety; initiatives such as the European Union's nuclear stress tests; and the review and, where necessary, the revision of the IAEA safety standards, and the continuous strengthening of the IAEA's peer review and advisory missions. Cooperation among international organizations, as well as a multitude of national and regional initiatives, has further strengthened nuclear safety globally.

The conference also examined ways in which nuclear safety can be further strengthened in the future with the advent of advanced technologies, small modular reactors, the use of the international legal instruments for nuclear safety and new ways to communicate, engage and build the trust of stakeholders, including the public. The Government of Japan has requested the IAEA's assistance to review the country's plans and activities regarding the safety aspects of discharging the treated water from the Fukushima Daiichi nuclear power plant into the sea.

The 689 conference participants, 131 speakers and panellists, and 216 observers from 68 countries and seven international organizations attended the conference either virtually or in person. Two keynote speeches and 105 presentations were given in the opening session and the five sessions and the 11 panel discussions that followed. The conference was held in cooperation with seven international organizations with valuable input from the Special Experts Committee.

This publication contains the Conference President's report, including summaries of the sessions and panels and the President's Call for Action, and summaries of the opening and closing remarks and the presentations from the sessions and panels. The annexes provide programme information.

The IAEA gratefully acknowledges the cooperation and support of the organizations and individuals involved in this conference. The IAEA officers responsible for this publication were G. Caruso of the Office of the Deputy Director General and T. Karseka Yanev of the Office of Safety and Security Coordination.

EDITORIAL NOTE

CONTENTS

1. INTRODUCTION

1.1. BACKGROUND

On 11 March 2011, Japan was shaken by what became known as the Great East Japan (Tohoku) Earthquake. It was followed by a tsunami which resulted in waves reaching heights of more than 10 meters. The combined impact and repercussions of the earthquake and tsunami caused great loss of life and widespread devastation in north-eastern Japan.

The IAEA's Incident and Emergency Centre (IEC) received information from the International Seismic Safety Centre at approximately 08:15 Vienna Time concerning a strong earthquake with a magnitude of 9.0 near the east coast of Honshu, Japan's main island. The subsequent tsunami triggered an accident at the Fukushima Daiichi Nuclear Power Plant (NPP), which was ultimately categorized as a Level 7 – Major Accident – on the International Nuclear and Radiological Event Scale.

In the initial days following the accident, the IAEA established teams to evaluate key nuclear safety elements and assess radiological levels. IAEA laboratories reviewed environmental data provided by the Japanese authorities on monitoring of the marine environment and also received terrestrial environment samples for independent analysis to examine and assess the radiation levels. The IAEA posted daily updates for its Member States and the general public on the IAEA website to provide information on the actions taken soon after the accident.

By September 2011, the IAEA developed an Action Plan on Nuclear Safety, endorsed by the IAEA Member States, which defined a programme of work to strengthen the global nuclear safety framework in response to the accident. In addition to the Action Plan on Nuclear Safety, a great deal of work has been conducted worldwide to strengthen nuclear safety. Through initiatives such as the European Stress Test, the adoption of the Vienna Declaration on Nuclear Safety in accordance with the objectives of the Convention on Nuclear Safety (CNS), as well as the multitude of national and regional initiatives, many safety improvements have been developed and implemented in recent years.

Work to implement the Action Plan on Nuclear Safety went on to form part of the 2015 Report by the IAEA Director General on the Fukushima Daiichi NPP Accident and its five accompanying Technical Volumes. They addressed the accident's causes and consequences and provided a comprehensive understanding of what happened and why, as well as the lessons learned. They consider the accident itself, emergency preparedness and response, radiological consequences of the accident, post-accident recovery, and the activities of the IAEA following the accident. Measures were taken, both in Japan and internationally.

The IAEA is continuously strengthening and increasing its peer review and advisory missions to Member States, which are conducted at their request. The IAEA safety standards have also been reviewed and, where appropriate, revised. All this, and several other measures such as the Action Plan on Nuclear Safety, are major contributions of the IAEA to further strengthening nuclear safety worldwide after the Fukushima Daiichi NPP accident.

After the announcement of its basic policy in April 2021 to discharge the treated water from the Fukushima Daiichi NPP into the sea, the Government of Japan requested assistance from the IAEA to review the country's plans and activities. The IAEA's assistance will address safety aspects of the handling of the water stored at the Fukushima Daiichi NPP – related to radiation safety of the public and the environment – as well as transparency. The review will be conducted against international safety standards, which constitute a global reference for protecting people and the environment and contribute to a harmonized high level of safety worldwide. On 8 July 2021, the IAEA and Japan agreed on the scope of technical assistance the Agency will provide. The signing of the Terms of Reference marks an important step as the document sets out the broad framework for how the IAEA will review Japan's plans and activities related to the water discharge.

The International Conference on A Decade of Progress after Fukushima Daiichi: Building on the Lessons Learned to Further Strengthen Nuclear Safety was organized by the IAEA and held in Vienna on 8–12 November 2021, reviewed lessons learned from the Fukushima Daiichi NPP accident and the actions taken worldwide in the 10 years after the accident to strengthen nuclear safety. This publication serves as proceedings material of the conference.

1.2. OBJECTIVES

The conference focused on two main objectives:

— Looking back on the lessons learned, experiences shared, results, and achievements from actions undertaken by national, regional, and international communities following the accident; and
— Identifying ways for further strengthening nuclear safety as part of the answer to maximizing the beneficial peaceful use of nuclear energy.

The conference gathered internationally recognized high level safety experts and other stakeholders to discuss initiatives taken in the aftermath of the accident, and initiatives for further strengthening nuclear safety, together with other aspects relevant to achieving the overall goal of helping to address the climate change crisis. Vital to this goal was the development of a conference President's action plan based on the conference.

1.3. KEY MESSAGES

The key messages for the conference were:

— Demonstration, not declaration, that we are safer now than ever before – thanks to the work done by the IAEA, its Member States, and other international organizations;
— There is no place for complacency – we have to maintain the momentum and continue strengthening nuclear safety;
— This is not a retrospective; it is an opportunity to plan for the future – it is 10 years on; we will apply the lessons learned from our achievements in future activities to enhance nuclear safety;
— The nuclear industry exists in a broader context – apart from technical matters, societal matters are equally important; in this regard, the matters of public trust, climate change, and the involvement of future generations, form a large part of global development and success.

1.4. STRUCTURE OF CONFERENCE AND THIS PUBLICATION

The conference was developed with the assistance of a Special Experts Committee, using a series of technical sessions and panels with invited speakers. The technical sessions provided specific success stories for the given topical areas and highlighted ongoing initiatives to strengthen nuclear safety. Discussion panels focused on cross-cutting strategic considerations, global challenges and actions to further enhance nuclear and radiation safety. The closing session provided an opportunity for the

President of the Conference to present a summary and conclusions of the conference, with visions and strategies for the future.

Sessions were opened by the Chairperson(s), followed by presentations by the invited speakers, then questions from participants and answers from the speakers. Panels were opened by the Moderator, followed by short opening statements from the panellists, then a discussion with conference participants.

The programme consisted of three main parts, after an Opening Session with keynote speakers that set the context for the conference. The three main parts were:

— Part I – International Perspective;
— Part II – Learning Lessons;
— Part III – Path Forward.

Part I, International Perspective, provided insights from various international organizations from their work in response to the accident, reflecting on their work, thinking about the future and demonstrating the global safety of nuclear energy and its robustness through its international institutions and systems.

Part II, Learning Lessons, involved high level policy and technical sessions on the overall themes central to the conference, and covered topics such as:

— Ensuring the safe generation of nuclear power;
— Emergency preparedness and response;
— Radiation safety;
— Post-accident recovery;
— International cooperation;
— Leadership and management for safety;
— Communication and trust building;
— International legal instruments for safety;
— Facing new challenges;
— Safety for nuclear power development.

This part was aimed at demonstrating the improvements made in nuclear safety but from a more technical and themed approach.

Part III, Path Forward, sought to identify actions to further enhance nuclear and radiation safety, considering cross-cutting strategic considerations and global challenges. It consisted mainly of dynamic panel discussions among leaders and experts from a variety of different perspectives.

The last panel of this part discussed the insights offered by speakers during the week, and conclusions and recommendations from the earlier sessions to identify and propose future actions that would be most beneficial in maintaining momentum.

In a closing session the President of the Conference presented the summary and conclusions of the conference, including visions and strategies for the future, and most importantly called for action.

In addition to these main parts of the conference, there were three special sessions covering topical subjects:

— Nuclear safeguards at Fukushima Daiichi NPP;
— Safety related aspects of Advanced Liquid Processing System (ALPS) treated waters after the accident; and vital to the future;
— Youth and the nuclear industry.

This publication starts with an executive summary, followed by the structure which reflects the conference programme to the large extent (see above). It also includes three annexes.

Annex I contains a detailed copy of the conference programme. Annex II contains the list of conference participants. Annex III lists the members of the Special Experts Committee, the Programme Committee and the IAEA Secretariat.

2. EXECUTIVE SUMMARY

The conference consisted of three main parts: an international organization perspective; learning lessons and the path forward, with three additional topical events: Safeguards at Fukushima, the Director General of the IAEA's session on ALPS treated water, and a special youth panel. There were technical sessions with question-and-answer periods, and moderated interactive panels followed by discussion, with invited speakers and panellists.

The Opening Session included welcoming remarks from Director General Mr R. M. Grossi, Ambassador of Japan Mr Takeshi Hikihara, Conference President Mr Mike Weightman, Scientific Secretary Mr Gustavo Caruso. The keynote speakers were Deputy Director General and Head of the Department of Nuclear Safety and Security Ms Lydie Evrard on the importance of international cooperation and the IAEA's support for Member States and Mr Hajimu Yamana of Japan Nuclear Damage Compensation and Decommissioning Facilitation Corporation with an overview from his experience on how to avoid nuclear accidents, grounded in the status of Fukushima Daiichi NPP post-accident.

Session A: Contribution of International Organizations to Global Safety, provided an opportunity for international organizations who contributed to global efforts during and after the accident to share their work and demonstrate that the peaceful uses of nuclear energy are safer now than ever before.

Session B: Ensuring the Safety of Nuclear Installations, summarized the lessons and associated actions from the accident and highlighted the different approaches and decisions that were taken to enhance nuclear safety and face emerging challenges.

Panel 1: Ensuring the Safety of Nuclear Installations – Minimizing the Possibility of Serious Off-Site Radioactive Releases, discussed measures that can be taken to ensure that serious accidents are very unlikely and highlighted actions to ensure that serious off-site radioactive releases will be avoided or minimized in line with the principles of the Vienna Declaration on Nuclear Safety. The panel also investigated the potential of advanced reactor technologies to practically eliminate the risk of off-site releases. Panellists discussed where a questioning attitude contributes to safety and where strict implementation is a prerequisite for the safe operation of nuclear installations.

In Session C: Preparing and Responding to a Potential Nuclear Emergency, speakers shared experiences on various aspects of managing emergency response, including protecting emergency workers and helpers and justifying protective actions.

Panel 2: Preparing and Responding to a Potential Nuclear Emergency – Robust Preparedness, discussed the importance of infrastructural elements for emergency preparedness and response, including regulatory requirements; clearly defined roles and responsibilities; pre-established plans and procedures; tools, equipment and facilities; training, drills and exercises; and a management system.

Session D: Protecting People Against Radiation Exposure, discussed challenges and successful approaches for protecting the public against radiation exposure and for ensuring timely and effective communication to the public.

Panel 3: Protecting People Against Radiation Exposure – Attributing Health Effects to Ionizing Radiation Exposure and Inferring Risks, discussed the perceived inconsistencies in the limitations for attributing radiation effects following low level radiation exposures vis-à-vis the latest estimates of

United Nations Scientific Committee on the Effects of Atomic Radiation (UNSCEAR) on attribution of effects and inference of risk, which became available after the Fukushima Daiichi NPP accident.

In Session E: Recovering from a Nuclear Emergency, speakers shared their experiences from the perspective of international organizations, government authorities, and local and regional leaders involved in the recovery operations from the Fukushima Daiichi NPP accident.

In Panel 4: International Cooperation, the panellists shared different perspectives on how international cooperation contributes to establishing an international framework and global commitment for nuclear safety.

The side event on Safeguards in a Challenging Location: How the IAEA Implements Nuclear Material Verification at Fukushima Daiichi NPP, discussed the safeguards activities undertaken by the IAEA at the site in the 10 years since the accident, the progress made in re-verifying the nuclear material that was left inaccessible by the accident, and the innovative technologies developed in response to the challenges on the site.

The Special Session by the Director General of the IAEA on Safety Related Aspects of Handling ALPS Treated Water at the Fukushima Daiichi NPP heard from key officials from the IAEA and Japan, about recent progress on the Government of Japan's plan to release ALPS (advanced liquid processing system) treated water from the Fukushima Daiichi NPP into the sea through controlled discharges, and shared information about future activities.

Panel 5: Youth and the Nuclear Industry, consisted of finalists in an IAEA essay competition on selected topics related to the themes of the conference for students and early career professionals. They discussed the themes and concepts from the winning essays and the critical role that the next generation will play in sustaining and ensuring a safe future for the peaceful uses of nuclear technology.

Panel 6: Safety for Nuclear Development, examined how vendors, recipients, and international organizations can all play a role in ensuring that nuclear safety remains a global priority and highlighted the role of a robust nuclear safety infrastructure in enabling future nuclear development.

Panel 7: Building Inclusive Safety Leadership, drew attention to the changing workforce demographics in operating organizations, regulatory bodies, and technical support and research organizations. Panellists discussed strategies to address potential challenges and how nuclear sector practices are keeping pace with other industries in having an inclusive workforce and safety leadership in the mid-21st century.

Panel 8: International Legal Instruments, discussed how the effectiveness of the international legal instruments for safety can be further enhanced.

Panel 9: Communication, Engagement and Trust Building, discussed how information can be shared accurately and in a timely manner in a way that is understandable to the target audience and builds trust with the public.

Panel 10: Facing New Challenges, discussed trends and global developments that have a direct impact on nuclear power in its current format. Opportunities to enhance safety under new circumstances were also discussed.

Panel 11: Call for Actions – Maintaining the Momentum, proposed actions for moving forward, using the key messages from the conference, coupled with the insights from speakers throughout the week to identify where future efforts and momentum will be most beneficial. The outcomes of this panel were incorporated into the President's summary and Call for Action.

The closing session included concluding remarks from IAEA's Director General Mr R. M. Grossi, Scientific Secretary Mr G. Caruso and Conference President Mr Weightman. Mr Weightman reflected on the information and experiences shared and introduced his President's Call to Action, a series of proposed actions under four headings: enhancing openness; embedding the lessons of Fukushima Daiichi NPP accident; better preparing for the wider use of safe nuclear power; and transferring knowledge for the future. This Call for Action is the major outcome of the conference, to aid the IAEA in planning for the next decade of safe nuclear power post Fukushima.

3. PRESIDENT'S REPORT

3.1. OBJECTIVE OF THE PRESIDENT'S REPORT

This report provides a summary of this important, timely and seminal conference hosted by the IAEA in cooperation with:

— The Food and Agriculture Organization of the United Nations (FAO);
— The International Labour Organization (ILO);
— The Organisation for Economic Co-operation and Development / Nuclear Energy Agency (OECD/NEA);
— The Preparatory Commission for the Comprehensive Nuclear-Test-Ban Treaty Organization (CTBTO);
— The United Nations Scientific Committee on the Effects of Atomic Radiation (UNSCEAR);
— The World Health Organization (WHO); and
— The World Meteorological Organization (WMO).

This report provides the background to the conference, its objectives, its structure, the conclusions and findings of the various sessions, and most importantly the President's "Call for Action" that was the principal output of the conference.

It was suggested that a mechanism needed to be put in place to ensure that the actions identified were taken forward and addressed adequately. The President, in his closing remarks, proposed that one possible way to address this would be via a suitable side event each year at the IAEA General Conference. Providing the President's "Call for Action" to other fora, such as the meetings of the Contracting Parties to the CNS and the IAEA review committees could also assist in its delivery.

3.2. THE PRESIDENT'S CALL FOR ACTION

During the conference, international experts recalled lessons learned, experiences shared, results, and achievements from actions undertaken by national, regional, and international communities following the Fukushima Daiichi NPP accident. Furthermore, they tried to identify initiatives for further strengthening nuclear safety in the context of the wider use of nuclear power to help combat climate change.

Nuclear power remains an important element of the global electricity production. Therefore, the overarching goal is: Safe nuclear power for all as part of the solution to climate change.

An important outcome of the conference was, based on lessons learned from the Fukushima Daiichi NPP accident and recognizing the global trends, a proposal of priorities for stakeholders regarding safe operation of nuclear power for the next decades. This specific outcome is the "President's Call for Action".

The aim of the "President's Call for Action" is to highlight that safe nuclear power exists in a broader context and that now is the time to plan for the next 10 years. Apart from technical matters, human, organizational and societal matters are equally important; and, the matters of public trust, climate change, and the involvement of future generations, form a large part of any global development and success.

The President's Call for Action falls into four areas:

1) Enhance Openness;
2) Embed Lessons from the Fukushima Daiichi NPP accident;
3) Better Prepare for the Wider Use of Nuclear Power;
4) Transfer of Knowledge for the Future.

To Enhance Openness:

— Summarize all the experience and work in response to the accident, as a foundation for the demonstration of the international institutions working together, as well as lessons learned globally and improvements made over the last ten years; and, as a basis for more effective working together in the future.

— Enable better understanding of the balance of radiological and non-radiological impacts through enhanced guidance on the justification and optimization of protective actions for decision makers. The guidance needs to be open and understandable to all stakeholders, including the public.

— Review the present system of radiation protection with regard to the inference of risks at low doses and the use of dose criteria when there are different balances to be made, to help facilitate better public understanding of the basis for decision making.

— Extend practical guidance in support of decision making on balancing risks associated with ionizing radiation vs other risks or benefits/detriments in policy, design, operations, decommissioning and waste management, providing examples for different circumstances.

— Promote and use relevant research findings on risk perception and human behaviour to develop further communication tools and guidance to help earn the trust of the public and stakeholders in the context of a global strategy.

— Foster the above-mentioned research and considering human behaviour in relation to radiation risks in communication tools and advice, so that policy makers and the public can make effective decisions during and after an accident.

— Reach out to the younger generations, especially leveraging such tools as the internet and social media for clear and concise communications that simplify technical concepts through analogies and illustrations to generate interest, understanding and earn trust.

To Embed Lessons from the Fukushima Daiichi NPP accident:

— Encourage Member States to timely implement any remaining safety improvements at existing nuclear power plants resulting from lessons from the Fukushima Daiichi NPP accident.

— Member States, vendors, licensees, regulators to ensure that new reactor technologies take into account lessons from the Fukushima Daiichi NPP accident by taking into consideration the IAEA safety standards.

— Relevant organizations to undertake a review to determine whether a more comprehensive view of public health consequences from an accident requires changes to guidelines for decision makers on protective actions following an accident. Criteria for the emergency phase and for the actions in the recovery phase need to be considered.

— Provide for the implementation of the UNSCEAR report on attribution of health effects and inference of risks by developing best practices and guidance for their application in radiation and nuclear safety, to provide better understanding on radiation effects and to avoid misconceptions among those responsible for making decisions on protective measures and members of the public.

— The IAEA Member States and relevant organizations to develop and promote strategies and share experiences of applying a systemic approach to safety related decisions, considering organizations, people, and technology, and their interactions to build robust safety systems.

To Better Prepare for the Wider Use of Nuclear Power:

— The IAEA to undertake a review and propose enhancements, as necessary, of the global system for nuclear safety (national frameworks, technical support capabilities, international cooperation), taking into account the potential for wider use of nuclear power in the future.

— Member States that are considering embarking on or expanding their capability in nuclear power to review their existing systems for nuclear safety.

— Develop strategies and build cooperation networks among reactor designers and operators to strengthen design and operational safety for preventing severe accidents and mitigating their consequences to avoid unacceptable radioactive releases, taking into account methods to determine how safe is safe enough using techniques such as cost–benefit analysis.

— Relevant international organizations to continue and bring to a conclusion the work on unified/harmonized standards and guidelines for radionuclides in commodities (food and others) that are based on risk.

— Member States to include generic planning for recovery and arrangements for recovery after an accident and establish the division of responsibilities between different parties in an emergency in the consultation process preceding the authorization of new facilities.

— The IAEA, OECD/NEA and Member States to elaborate the concept of "Culture for Safety" taking into consideration national factors.

— Member States to consider a national approach for the implementation of a robust culture for safety at all levels from staff to policy makers.

— Develop further guidance on promoting and implementing leadership approaches to guard against complacency, continuously seeking innovation and improving nuclear safety.

— Member States to ensure emergency management systems are established under an "all hazards" model, optimizing resources and integrating various functions, to enable a holistic approach to responding to different hazards at the same time.

To Transfer Knowledge for the Future:

— The IAEA to enhance facilitation of the implementation of knowledge transfer strategies to countries embarking on nuclear power programmes.

— The IAEA to consolidate the international experience of recovery from accidents and review their impact in order to take better based decisions on emergency planning and recovery.

— Ensure the systematic application of knowledge management to the decommissioning and recovery from the Fukushima Daiichi NPP accident to facilitate decommissioning of redundant nuclear sites and be better prepared to respond to and recover from any potential nuclear accident.

— The IAEA and Member States to increase capacity and facilitate capability building to ensure the availability of regulatory and technical competence for the safety of new nuclear facilities.

— The IAEA to establish a sustainable and open forum for sharing and promoting best practices in the remediation and decommissioning of nuclear facilities.

— The IAEA and Member States to promote and facilitate the scientific understanding of ionizing radiation and the peaceful use of nuclear energy in general education programmes for the youth of tomorrow to inform future generations and encourage them to pursue careers in the sector.

3.3. PRESIDENT'S SUMMARY OF THE CONFERENCE

Opening Remarks

R. M. Grossi welcomed the participants, noting that nuclear safety always comes first, and nuclear energy has to form part of the energy mix to combat climate change. He stressed that this conference was one way to demonstrate that, as a global community, the society has taken stock, learned the lessons of the Fukushima Daiichi NPP accident and applied them. However, one is also looking forward, continuously improving and learning, never complacent, innovating and adapting to meet the challenges of the future. He expressed the need to communicate all of this clearly and with transparency so that governments and the public can trust that nuclear energy is safe and make informed decisions about its use in a de-carbonized world.

T. Hikihara expressed sincere appreciation for the conference and noted the substantial improvements in nuclear safety over the past 10 years. He briefly explained Japan's efforts to date and their plans going forward, including sweeping reforms so that the Fukushima Daiichi NPP accident will never happen again.

These include:

— Ongoing decommissioning of the NPPs at the Fukushima Daiichi site;
— Sharing information with other States and CNS signatories;
— Actions for environmental recovery of the area;
— Continuing efforts to enhance Japan's nuclear safety;
— Working with the IAEA and cooperating with the international community, including training at the Fukushima Daiichi site for newcomer countries.

M. Weightman welcomed the participants to the conference and gave an overview of the programme. He expressed his hope that, by demonstrating that the nuclear industry has learned the lessons from the Fukushima Daiichi NPP accident, it has made real progress over the last ten years and is continuously improving nuclear safety; with more plans for the future, we can show that safe nuclear power can be an important part of the solution to the climate change crisis, providing clean energy for all.

G. Caruso noted that over 700 people were attending the conference, whether virtually or in person. He emphasized the importance of open and transparent discussions and the need to hear the views of all participants. The key message of the conference is that nuclear safety has a role to play at all levels and nuclear energy is of vital importance to the future.

Keynotes

L. Evrard noted that international cooperation is essential, as nuclear safety is a national responsibility, but accidents affect us all globally. She expressed the need to learn in a holistic manner over many dimensions, as demonstrated by the conference programme. Cumulative experience is invaluable and experiences from more mature nuclear countries can be lessons to newcomers. She outlined the IAEA's support for Member States and reiterated that we need to continuously identify ways to further improve nuclear safety. This conference would assist in that regard, shaping IAEA activities for the next decade.

H. Yamana gave an overview from his experience on how to avoid nuclear accidents, grounded in the status of Fukushima Daiichi NPP post-accident. He noted six areas where lessons were learned:

— Institutional issues, such as safety culture and regulatory independence;
— Technological issues, with defence in depth and diversity of safety systems being key elements;
— Evacuation considerations and the need to pay more attention to its psychological and social effects;
— Accident management and emergency preparedness and response, with staff capability and collaboration with government key to dealing with an emergency;
— Social and environmental recovery, including compensation for survivors and damage; and the challenge of restoring a contaminated environment;
— Decommissioning of the site with emphasis on the importance of stakeholder involvement in the final solution to decommissioning, and the consideration of the serious reputational damage caused by the accident – the impacts are global and regaining the public's trust is a huge challenge.

He encouraged participants to work together to establish better and safer systems for nuclear safety moving forward with the lessons learned from Fukushima Daiichi NPP accident.

Session A – Contribution of International Organizations to Global Safety

Speakers in this session noted the activities of their various organizations after the Fukushima Daiichi NPP accident, with several cross-cutting themes and lessons becoming evident:

W. Magwood of OECD/NEA noted that:

— The NEA issued a report shortly after the accident that stated that NPPs are safe, but they need to be made more resilient to very extreme threats.
— Safety culture is key to success. The NEA reported five years ago on changes implemented globally; training, equipment and procedures have all changed in NPPs.
— The removal of the damaged core from the Fukushima Daiichi NPP is a huge challenge and the international community has to help.
— Important lessons learned included:
  • The ability to recover is as important as avoidance;
  • Safety culture is as important as technical expertise;
  • Improving regulatory authority, recognizing the health aspects, and seizing the opportunity for economic development are important in recovering from a nuclear accident;
  • Engagement of the public and other stakeholders is crucial. Nuclear power is essential to the future, to the environment, the economy and civilization, however nuclear power cannot play a strong role if the public does not believe it is safe. Advanced technologies can help such as small modular reactors (SMRs), but the public needs to be part of the discussion.

I. Engkvist of WANO noted that:

— While operators are responsible for safety, regulators, governments and organizations contribute to safety and all need to work together.
— All aspects of operating NPPs have been strengthened in the past 10 years. Institutions have been strengthened and operators are humbler and wiser, leading to even safer NPPs.
— There are programmes and processes in place to identify operational weaknesses to enhance safety.
— Leadership is key and efforts are being made in this area as well.
— WANO recognizes the value of nuclear energy as a reliable, safe and economically viable source of power. The business and political pressures, geographic challenges for global cooperation and collaboration are unique to this industry but we can be safe and productive in the future.

G. Caruso gave the IAEA perspective on the accident, with an overview of the actions taken at the time and over the past 10 years, including:

— The development and implementation of the IAEA Action Plan on Nuclear Safety;
— Numerous international meetings and documents on analysis and lessons learned from the accident, concurrent with IAEA support to Member States to enhance nuclear safety in peer reviews and advisory missions;
— The review of the IAEA safety standards to incorporate findings and lessons from the Fukushima Daiichi NPP accident;
— Going forward, the IAEA has also been requested to oversee the discharge of ALPS treated water into the sea and continues to provide support to the Fukushima Prefecture.

G. Hirth of UNSCEAR presented information on the UNSCEAR 2020 Report on Fukushima[1]. She noted that, overall, the findings of the UNSCEAR 2020 Report are generally consistent with those in

---

[1] UNSCEAR 2020/2021 Report Volume II: "Sourses, Effects and Risks of Ionizing Radiation" Annex B: Levels and effects of radiation exposure due to the accident at the Fukushima Daiichi Nuclear Power Station: implications of information published since the UNSCEAR 2013 Report

the UNSCEAR 2013 Report but there is now more information available to support the Committee's conclusions, which are:

— The accident led to no adverse documented public health effects that were directly attributable to radiation exposure from the accident;
— Future cancer rates that could be inferred from radiation exposure from this accident are unlikely to be discernible;
— Increased incidence of thyroid cancer observed in children in Japan was judged to be the result of extensive ultra-sensitive screening;
— Lessons learned included:
  • The importance of gathering quality measurement data taken as soon as possible during and after an accident;
  • The need to understand the base rate of cancer and to follow up with estimates due to radiation effects;
  • The need to continue high quality research and base information on science.

G. Graham of CTBTO presented an overview of key achievements in the CTBTO radionuclide monitoring technology over the last decade, in terms of additional station certification and novel software tools for data analysis and dissemination, including:

— In March 2012, the CTBTO became a member of the Inter-Agency Committee on Radiological and Nuclear Emergencies (IACRNE). Further civil applications for disaster risk reduction have been proposed.
— According to the Joint Radiation Emergency Management Plan, the critical response tasks of the CTBTO during an emergency phase is to provide real-time particulate and noble gas monitoring data including confirmation of no detection.

He noted the unique capability of inductively coupled plasma mass spectrometry to help in looking at the consequences of nuclear accidents, and that the accident changed the way analysis is done.

C. Blackburn and G. Dercon of the Joint FAO/IAEA Centre of Nuclear Technologies in Food and Agriculture gave a presentation of activities in their area from the lessons learned, covering:

— Improved international standards and guidance for radionuclides in food and agriculture;
— Targeted technical aids for use by agricultural departments in general;
— Coordinated international research activities to improve and extend remedial options for agricultural land in less well studied agricultural domains.

J. Pintado Nunes of ILO outlined some of the ILO's work on occupational safety and health and radiation protection since the Fukushima Daiichi NPP accident, including:

— The declaration of occupational safety and health as fundamental to decent work in the ILO Centenary Declaration 2019, with a resolution requesting study of the options for it to be incorporated as a possible fifth category of ILO Principles and Fundamental Rights at Work.
— ILO's Global Call to Action for human-centered recovery from COVID-19.
— ILO's international legal framework, including the Radiation Protection Convention, 1960 (No. 115), concerning all activities where workers may be exposed to ionizing radiation, with 50 ratifications, and its accompanying Recommendation (No. 114), that specifically concerns

the protection of workers against ionizing radiation. Member States report periodically on their adherence to the Convention.

— Observations on Convention No. 115 in 2015 noted the need for emergency preparedness and response plans that optimized protection strategies with reference levels in emergencies within or preferably below the range of 20–100 mSv, with measures taken to ensure emergency workers are not subject to exposure of more than 50 mSv.

— Development and promotion of a training package for IAEA Safety Standards Series No. GSG-7, Occupational Radiation Protection.

— International cooperation in the work of several platforms with the IAEA, OECD/NEA, Inter-Agency Committees and UNSCEAR on occupational radiation protection.

L.P. Riishojgaard presented on the capacity of the WMO to assist in nuclear accidents, noting that:

— They have specialized stations for monitoring radioactive emissions in the atmosphere to respond to nuclear emergencies.

— They now cover all environmental data, including climate, hydrology, oceans, etc.

— In October 2021, they updated their policies and vehicles to finance data sharing, particularly with developing countries.

— They aim at uniform coverage both surface and satellite with a global data processing and forecasting system for a global modelling network that makes products and services available to countries and international organizations such as the IAEA.

M. Neira of the WHO stated that:

— The Fukushima Daiichi NPP accident transformed the local, national and global response to nuclear accidents.

— International health regulations for managing health emergencies require coordination and cooperation, so that roles are complementary, and duplication is avoided.

— Non-radiological aspects can have terrible effects, with the social fabric destroyed and people deprived of privacy and access to health care. The mental health and psychosocial and health determinants should be taken into account in preventative measures, as they can last for decades.

— There is a need to invest in environmental and ecological protection and planning for the future.

Session B – Ensuring the Safety of Nuclear Installations

As a result of the Fukushima Daiichi NPP accident, the importance of continuously challenging the existing assumptions regarding nuclear safety to prevent future accidents was emphasized by the conference. Countries reviewed and reinforced, as necessary, the capability of nuclear installations to withstand or control possible accidents originating from extreme conditions and/or extreme external events to minimize risk. The IAEA and the Member States reviewed and revised their safety frameworks, including updates to the IAEA safety standards, to enhance nuclear safety at the national level and worldwide.

During the session, speakers summarized the lessons and associated actions from the accident and highlighted the different approaches and decisions that were taken to enhance nuclear safety and face emerging challenges.

The opening to this session consisted of a series of speakers, followed by a high level discussion of measures to ensure safety. It was followed by a specialist panel discussion on minimizing the possibility of serious off-site releases.

K. Watanabe looked back at 10 years of change and improvements in Japan in light of the lessons learned from the Fukushima Daiichi NPP accident, including:

— Reform of the regulatory system, both the regulator and the regulations;
— Establishment of the Nuclear Regulation Authority (NRA) as an independent regulatory body, with the separation of regulation and promotion of nuclear energy;
— New regulations on severe accidents that add more measures against severe external events with enhanced safety margins for tsunamis and earthquakes;
— An environment that encourages responsibility, open and frank discussion, a questioning and learning attitude and continuous improvement.

He noted that no one died from exposure to radiation, but many thousands remain evacuated and work to decommission and decontaminate is still ongoing.

F. Aparkin provided an overview of the Russian developments for nuclear power, covering:

— NPP safety assessments in light of the lessons learned from the accident at the Fukushima Daiichi NPP;
— Building a strengthened safety concept for NPPs with new generation VVER reactors, taking into account national experience and lessons learned from the accident at the Fukushima Daiichi NPP;
— Building a safety concept for prospective NPPs using the examples of land based stationary NPPs and floating nuclear power units, taking into account national experience and lessons learned from the accident at the Fukushima Daiichi NPP;
— The review and update of the IAEA safety standards in the light of the developments in nuclear safety.

P. Tippana reported on changes at STUK in the area of safety culture as a result of the Fukushima Daiichi NPP accident. STUK developed a safety culture assessment programme in response to Japan's acknowledgement of the influence of culture on the disaster. He noted the following findings:

— Organizational culture has an effect on nuclear safety. We need to pay attention to the effect of various levels of culture, including national culture, on safety culture.
— Survey results revealed that the Finnish system generally supports a good, healthy, safety culture. However, they did find that the same cultural attributes, on their own or in their composition can have both positive and negative impacts on safety. STUK is incorporating these findings into its operations.
— Regulatory oversight culture has an impact on the operators' safety culture.
— Societal expectations change; they can focus on confirming compliance with regulations or on client driven safety performance.

A. Pelle offered a review of changes made at EDF as a result of the accident, noting that they do not define reactor life. Previously they had 10-year inspections with a compliance review. After the Fukushima Daiichi NPP accident, they implemented the following measures to increase the robustness of NPPs:

- Safety improvements for station blackout and external hazards, constructing new equipment on each reactor and making major improvements to crisis organization and training;
- The creation of the Nuclear Rapid Action Force (known by the French acronym 'FARN'), a response task force with the objective to access all sites within 12 hours and be operational within 24 hours;
- Increased site autonomy to three days with changes to primary circuit supply, reactor buildings, fuel buildings, spent fuel pools and mitigation of core meltdown in severe accidents;
- Improved core meltdown prevention;
- There has been a paradigm shift from preparing for specific hazards to preparing for all hazards after the Fukushima Daiichi NPP accident, with accompanying changes in safety culture, operations and regulatory oversight.

The question and answer (Q&A) discussion highlighted the following:

- The need for a strong safety culture to maintain public trust in decision making;
- The need to employ a systemic institutional approach in every decision to ensure safety at different stages in the lifetime of NPPs;
- The need to consider the issue of "how safe is safe enough?" and accident preparedness vs prevention.

Panel 1 – Minimizing the Possibility of Serious Off-Site Radioactive Releases

Findings from the Fukushima Daiichi NPP accident have shown that a drive towards continuous safety improvement leaves no place for complacency. Technical, organizational and regulatory measures taken to enhance safety further reduce the likelihood of occurrence of accidents with serious radioactive releases. On the other hand, as such an accident might still occur, the nuclear industry and regulatory bodies need to be prepared for the unexpected.

The panel discussed measures that can be taken to ensure that serious accidents are very unlikely and to highlight actions to ensure that serious off-site radioactive releases will be avoided or minimized, in line with the principles of the Vienna Declaration on Nuclear Safety. This included specific examples from different countries explaining the 'why' as well as the 'what.' The most significant changes noted were:

- An updated approach to safety analysis and hazard assessment programmes;
- Examining 'black swan' events and giving operators the resources to deal with them;
- Changing the regulatory mindset, realizing that a lot of what used to be impossible is possible and imagining new situations;
- Recognizing the need to be open minded and reviewing safety culture and leadership through this lens.

The panel also investigated the potential and limitations of advanced reactor technologies to practically eliminate the risk of off-site releases. They discussed the innovative passive systems in SMRs, the designs of new reactors that consider the issues with current reactors, the need to engage with the public in a two-way dialogue and communicate clearly what is safe and what is not safe. Panellists discussed the importance of leadership for developing cultures for safety and where a questioning attitude contributes to safety and where strict implementation and following instructions is a prerequisite for the safe operation of nuclear installations. A questioning attitude and continual review are necessary to prevent further accidents; however, it is important to avoid endless analysis that results in no progress. A strong regulator that can consider a multitude of perspectives in decision making and maintain public trust is key.

Session C – Preparing and Responding to a Potential Nuclear Emergency

An integrated and coordinated emergency management system for preparedness and response for a nuclear emergency has to be in place at the national level for effective action. The Fukushima Daiichi NPP accident has shown that these arrangements have to cover, among other things, the case of responding simultaneously to a nuclear emergency and a natural disaster. In this session, experiences were shared on various aspects of managing emergency response, including protecting emergency workers and helpers and justifying protective actions.

Findings included the following:

— In a nuclear emergency, protective actions have to be justified and implemented in an effective and timely manner, doing more good than harm.
— A comprehensive approach to planning and decision making has to be followed to ensure balance between potential radiological consequences, non-radiological consequences, and health hazards, with special consideration to sensitive population groups, such as children, pregnant women, and those in nursing homes and chronic care facilities.
— Communicating with and educating the public is vital in the preparedness phase.
— International exercises like ConvEx-3 are important to test and build capacity at the international level.
— Capabilities should be integrated into a plan for all hazards.
— IAEA safety requirements and generic criteria address the termination of a nuclear emergency and the subsequent transition to an existing exposure situation; however, the Fukushima Daiichi NPP accident highlighted that further guidance was needed.
— The Emergency Preparedness and Response Standards Committee (EPReSC) was established as a direct result of the Fukushima Daiichi NPP accident. It developed a road map for the future, driven by Member States' needs, a gap analysis and mapping, including protective strategies for planned, emergency and existing stages. Governments have suggested that this guidance should be elevated to safety standard status in order to be fully utilized.
— Emergency preparedness is more than just the plan: practice is key. Therefore, workers need to be designated in advance and given instructions and duties before an emergency. They need to participate in practice sessions that simulate real world conditions.

A specialist panel discussed how to ensure emergency arrangements are robust, in light of the lessons learned.

Panel 2 – Preparing and Responding to a Potential Nuclear Emergency – Robust Preparedness Arrangements

To be robust, preparedness arrangements need to be able to respond to an emergency at an NPP that might occur simultaneously with a natural disaster. The response to a nuclear emergency involves many national organizations, as well as international organizations and, therefore, has to be coordinated and effective. This panel discussed the importance of infrastructural elements for emergency preparedness and response, including regulatory requirements; clearly defined roles and responsibilities; pre-established plans and procedures; tools, equipment and facilities; training, drills and exercises; and a management system.

Conclusions included:

— Effective communication, talking with and the inclusion of stakeholders, including non-governmental organizations (NGOs) and the public, is needed in the preparedness phase.
— Simulation exercises should be done with people who will participate in the emergency response, including local public safety officials and government representatives, as behaviours and risk perception change when in an actual emergency.
— A multidisciplinary approach to emergency preparedness is necessary.
— It is crucial to build trust and engagement prior to an emergency.
— Need to communicate the benefits and safety of nuclear power to help the world address climate change and de-carbonized energy, including the value of SMRs and advanced reactors and their need for less extensive emergency planning arrangements and less constrictive siting criteria.
— The media is the most direct route to the public; however, an authoritative source is necessary given the dissemination of misleading or incorrect information in the democratization of information. The IAEA is one of the best sources for information on nuclear safety.

Session D – Protecting People Against Radiation Exposure

An important lesson from the Fukushima Daiichi NPP accident is the difficulty that non-specialists have in understanding the international system of radiation safety, including the principles and criteria for radiation protection. It is important to communicate the rationale behind the judgement as to whether and how radiation doses to the public should be averted, and to make clear that justification of protective measures and actions is based not solely on radiation science but on consideration of the overall benefits and detriments to society and the individual. Furthermore, guidance on monitoring doses to the public in the aftermath of an accident can be limited and this potential lack of information might create public anxiety. The public is particularly concerned about the protection of children and pregnant women after a nuclear accident. This session discussed such challenges, and successful approaches for protecting the public against radiation exposure while ensuring timely and effective communication to the public.

A high level discussion covered some of the main issues in this area, followed by a panel discussion addressing the particular aspect of the risk of low dose radiation exposure.

Session speakers noted the following:

— It has been shown that there would be no discernible health effects from radiation from the Fukushima Daiichi NPP accident over time; however, the public remain concerned. Care needs to be practiced in attributing health effects to radiation when there are none observed, or they can only be inferred.
— There were significant non-radiological (psycho-social) impacts which need to be recognized. However, there is no 'one-size-fits-all' approach to deal with the mental well-being and psychosocial impacts of an emergency.
— Potential non-radiological consequences need to be taken into consideration when implementing protective actions.

— It is important to take an all-hazards approach, considering both radiological and non-radiological consequences and the emotional dimension of risk perception, and practical tools need to be developed and made available to decision makers.

— The scientific community needs effective communication methods to convey scientific facts and evidence in a way that can be easily understood and trusted by the public, particularly as there are other sources exposing the public to misleading information.

Panel 3 – Protecting People Against Radiation Exposure – Attributing Health Effects to Ionizing Radiation Exposure and Inferring Risks

In the aftermath of the Fukushima Daiichi NPP accident, the radiation risk estimates used for radiation protection purposes were misinterpreted by the media and members of the public. While such estimates are intended for inferring risks based on assumptions, they were used to project absolute numbers of radiation-induced cancers following low-dose radiation exposure resulting from the accident. This resulted in disproportionate perceptions of risks by members of the public and might have contributed to increasing public anxiety with its associated health detriments.

The limitations of epidemiological studies for attributing radiation effects following low level radiation exposures need to be discussed and clearly explained. One aspect covered by this panel was the validity of the linear no-threshold theory in the light of the current understanding of radiation-induced health effects.

The panellists agreed that:

— The linear no-threshold (LNT) approach is a practical and simple model for radiation protection purposes; however, it remains an unproven model for the interpretation of the radiation effects at low doses, and great care needs to be taken in trying to apply it to low doses for which an increase in cancer risk is deemed not proven. Indeed, the statistical power of epidemiological studies is too low at such doses to support firm conclusions and where radiation biology studies provide insufficient evidence for either LNT or other general dose–response models.

— What is needed is a distinction between the attribution of health effects and the inference of risks.

— Epidemiological studies alone cannot elucidate health effects of radiation at low doses.

The panellists discussed the question of whether accurate estimates of the effect of low doses of radiation are necessary, since the risk of health effects can be negligible, and with the uncertainties in estimating these effects, incorrect interpretation can have a detrimental impact on decision making. In terms of public communication, it is important to answer the questions asked in an easily understandable way, enabling the public to make risk informed decisions.

Session E – Recovering from a Nuclear Emergency

The nuclear industry exists in a broader context; therefore, recovering from a nuclear emergency is a complex social, political, economic, technical, and scientific process. It requires coordination among a wide range of stakeholders and consideration of many aspects. This session covered several related topics, including: the role of technology and innovation; the involvement of the public in decisions on remediation efforts; and the identification of challenges that can inform future planning.

In this session, speakers shared their experiences from the perspective of international organizations, government authorities, and local and regional leaders involved in the recovery operations from the Fukushima Daiichi NPP accident. Points made included:

— Planning for recovery is part of assuring overall safety and emergency preparedness over the entire lifetime of a facility.
— The Fukushima Daiichi NPP is now at the beginning of the decommissioning phase, 10 years after the accident, with a roadmap expected to take 30–40 years. This is done in parallel with ongoing off-site reconstruction.
— There are challenges in contaminated water management, fuel and fuel debris removal, and waste management.
— The end state for the site and area is not yet decided and requires the involvement of local stakeholders to reach an agreed outcome.
— There are a wide variety of health issues associated with people in contaminated areas, including those evacuated. The Fukushima Health Management Survey after the accident showed psychological distress and an increase in lifestyle diseases, due not to radiation but to the effects of evacuation and reduced accessibility to health care. Stigmatization from the media also led to negative health impacts.
— As decontamination efforts continue, there is a need for further international cooperation to clarify and harmonize criteria for radioactivity in commodities and consumer goods.

Panel 4 – International Cooperation

The Fukushima Daiichi NPP accident emphasized the importance of international cooperation in safety related areas, including safe operation, emergency preparedness and response and regulatory effectiveness, and of incorporating lessons from the accident into national programmes to build capacity for more resilient systems.

Institutional networks for safety, such as regional networks, knowledge networks and regulatory forums, provide a platform for information exchange and help to optimize resources, compare processes, procedures and policies, identify good practices, identify and address existing gaps and needs.

The panellists shared different perspectives on how international cooperation contributes to establishing an international framework and global commitment for nuclear safety, and what more could be done. Findings included:

— International, regional and bilateral cooperation is vital for establishing mechanisms for provision of assistance and it is important to avoid duplication of efforts.
— Cooperation mechanisms established and agreed in case of emergency need to be in place for strengthening nuclear safety worldwide.
— International legal instruments and IAEA safety standards evolved after both the Chernobyl and Fukushima Daiichi NPP accidents to reflect findings and streamline the direction for international cooperation.
— The IAEA plays a vital role in facilitating the independent and transparent exchange of experience, where all parties benefit. Cooperation in such areas as post-accident recovery, leadership, safety culture and capacity building has been a focus.
— In nuclear safety, technical cooperation should be utilized for both developed and developing countries.

— International cooperation is needed and will be instrumental in avoiding the politicizing of nuclear safety.

Panel 5 – Special Topical Event – Youth and the Nuclear Industry

The IAEA invited students and early career professionals up to 30 years of age to submit essays on selected topics related to the themes of the conference. The aim of the essay competition was to promote creative and innovative thinking and highlight the critical role that the next generation will play in sustaining and ensuring a safe future for the peaceful uses of nuclear technology in areas such as nuclear power, food and agriculture, water management, and human health. The essay competition attracted 250 submissions from 60 countries. Finalists were selected through a blind evaluation process and attended the conference and participated in this special youth panel.

Discussions covered several topics including:

— The future of nuclear power generation, including the need for the timely implementation of safety standards to ensure that future generations remain safe in the long run, and that maintaining safety stays as the highest priority.
— Recognizing that nuclear power is even safer, the inclusion of youth in the nuclear industry and their innovative ideas in decision making will help drive it forward further;
— The importance of speaking the public's language, simplifying technical conceptualizations through analogies and illustrations in developing public understanding, interest and trust in the safety and reliability of nuclear power;
— Tools such as the internet and social media for clear and concise communication (and public relations) with stakeholders especially young people;
— Including the subject of nuclear power as part of education programmes to inform future generations and encourage them to pursue a career in the nuclear field;
— The importance of non-nuclear power applications;
— The use of the nuclear industry as a tool for development worldwide and the effect of future technologies such as SMRs.

Special Topical Event – Performing Safeguards at the Fukushima Daiichi NPP Site

Through a set of technical measures, the IAEA verifies that States are honouring their international legal obligations to use nuclear material and technology for peaceful purposes only.

To support the drawing of safeguards conclusions, the IAEA Department of Safeguards relies heavily on the timely access of inspectors to nuclear material and facilities. The conditions on the Fukushima Daiichi NPP site after the impact of the 2011 Great Tohoku Earthquake and tsunami posed unprecedented and continuously evolving challenges in fulfilling this mandate.

This side event discussed the safeguards activities undertaken by the IAEA at the site in the 10 years since the accident, from the post-accident response until the current situation, the progress made in re-verifying the nuclear material that was left inaccessible by the accident, and the innovative technologies developed in response to the challenges on the site, such as verification for damaged structures. The challenge for the future is the transition from a post-accident site to a decommissioning site (including the new buildings).

Special Topical Event – Safety Related Aspects of Advanced Liquid Processing System (ALPS) Treated Waters after the Fukushima Daiichi Nuclear Power Plant Accident

In April 2021, Japan announced the basic policy on handling of the treated water stored at the Fukushima Daiichi NPP, which is to discharge the treated water into the sea surrounding the plant, subject to domestic regulatory approvals. Soon after, the Japanese authorities requested assistance from the IAEA to monitor and review those plans and activities related to the discharge of the treated water to ensure they will be implemented in a safe and transparent way. Rafael Mariano Grossi, Director General of the IAEA, committed to provide support to the Government of Japan before, during, and after the water discharge, and the IAEA Secretariat has started initial planning and implementation activities associated with its review. A Task Force, comprised of the IAEA Secretariat as well as international experts, has been established and reviewing the Government of Japan's activities related to the treated water discharge.

The purpose of this Special Event was to hear from key officials from the IAEA and Japan, about recent progress and to share information about future activities. The main aspects covered were:

— The discharge of ALPS treated water will begin approximately two years after the announcement of the basic policy. Before and after the start of the discharge, enhanced marine monitoring will be conducted and made public by Japan in a thorough and transparent manner. This is based on over six years of technical discussions with continuous stakeholder involvement and communication with the international community.
— The Agency has been reviewing Japan's plans with respect to ALPS treated water using international safety standards as a benchmark; the review will occur before, during and after the planned discharge with components covering safety assessment, regulatory activities, sampling, and environmental monitoring (including corroboration of key data). Outputs of the IAEA review will include progress reports and updates to the public and international community at conferences, and briefings to Member States. Additionally, the IAEA will also conduct independent source and environmental monitoring to corroborate the data published by the Government of Japan. This work will be done in collaboration with IAEA and third-party laboratories.
— IAEA will publish the results of these reviews and distribute them amongst stakeholders, as appropriate. A website dedicated to this project has been established by the IAEA so that all stakeholders, including the public, have access to timely information.

Panel 6 – Safety for Nuclear Development

The IAEA and Member States have a joint responsibility to ensure that the sharing of nuclear technology is executed in a way that adheres to the highest standards of nuclear safety, security, and non-proliferation. Currently, several countries are embarking on new nuclear power programmes, while others are expanding their existing uses of nuclear and/or radioactive material for industrial, medical, and research purposes.

This panel examined how vendors, recipients, and international organizations can all play a role in the system for ensuring that nuclear safety remains a global priority,and highlighted the role of a robust nuclear safety infrastructure and system in enabling future nuclear development, building on INSAG-27, Ensuring Robust National Nuclear Safety Systems – Institutional Strength in Depth.

Panellists agreed that safety is a prerequisite to nuclear power generation and noted that:

— Multilateral cooperation and collaboration are vital.
— Regulatory independence is key, and development of national capacity is important as more and more countries develop nuclear power programmes.

— Safety is the responsibility of the operator, but the regulatory body plays a vital role in ensuring that there is adequate financial support, and that competence is developed and maintained.
— There is a need for appropriate regulatory structures, institutional and human resource capacities to be developed, using IAEA safety standards and advisory missions, to strengthen national systems for safety.
— National legal and regulatory frameworks need integrated management systems, and knowledge management and transfer are vital as technical support organizations (TSOs) and vendors assist in capacity building.
— Design and operating experience are to be shared.
— Responsible vendors will build national responsibility and provide support for the entire lifecycle of the facility and provide for the orderly transfer of intellectual property if and where needed.
— Safety has to be built into the supply chain.
— The current framework needs to be re-examined for sustainability. with the advent of SMRs and advanced reactor technologies, international and national frameworks for safety will have to evolve.

Panellists also discussed the international potential for mutual recognition of safety assessments, as well as the concept of an international TSO created under the auspices of the IAEA.

Panel 7 – Building Inclusive Safety Leadership

Global and national labour markets are changing. Personnel with different backgrounds, attitudes, expectations and competencies are joining the nuclear sector. The purpose of the panel was to draw attention to the changing workforce demographics in operating organizations, regulatory bodies, TSOs and research organizations.

Panellists discussed strategies to address potential challenges and exchanged ideas on how nuclear sector practices keep pace with other industries to have an inclusive workforce and safety leadership in the mid-21st century. These included:

— Guarding against complacency, being open to other perspectives;
— Retaining strong government support for nuclear safety leadership;
— Setting example of safety culture for operators by regulators;
— Maintaining a consistent approach to safety culture throughout the supply chain and across all stakeholders;
— Using an integrated management system to integrate multi-cultural workforces under one cohesive organization;
— Recognizing that a listening attitude is as important as a questioning attitude in leadership;
— Harmonizing safety culture concepts, working towards making nuclear safety culture universal, with one interpretation, to ensure important messages across cultural barriers. The IAEA can help by providing tools and advisory services;
— Learning from other industries, such as incorporating a mix of reflexive and reflected learning as in the aviation industry;
— Reducing uncertainty and training staff for other opportunities, e.g. decommissioning, in long term and phase out operations.

The panellists recognized that:

— Leadership and culture are equally as important to nuclear safety as technical aspects, such as defence in depth.
— Safety has to be the overriding priority, with leadership providing clear vision, values and expectations and following up to enforce those expectations, in order to establish a strong foundation for a culture for safety.
— Knowledge management and knowledge transfer are crucial to an inclusive workforce, in tandem with a leadership focus on succession planning and retaining talent.

Panel 8 – International Legal Instruments

Over the past four decades, several important international conventions and other international legal instruments have been adopted and progressively strengthened to achieve and maintain a high level of nuclear safety worldwide. International conventions are complemented by national policies for safety and regional agreements. Other initiatives also form part of the ongoing international effort to strengthen nuclear safety, such as the Vienna Declaration on Nuclear Safety that was unanimously adopted by the Contracting Parties to the CNS in 2015. The IAEA plays a critical role in maintaining the international legal framework for nuclear safety. The establishment of the IAEA safety standards through an international consensus process assists in the harmonization of nuclear regulations and helps States to comply with existing international legal instruments.

This panel discussed how the effectiveness of the international legal instruments for safety can be further enhanced. Panellists noted that:

— Safety is the responsibility of the operator, but conventions allow us to check on progress in lessons learned, as nuclear safety is always a work in progress. Contracting parties of conventions perform a peer review process.
— It is important to note that governments sign onto these instruments, not operators or regulators. The conventions are not applied directly but are reflected within national laws.

Suggested enhancements included:

— Using more modern communication tools to allow greater participation without the need to be present in Vienna (or another centre) might increase the effectiveness of existing legal instruments;
— Reviewing what the benefits to countries should be, especially the smaller ones as it is very intimidating for them to compile national reports;
— Using different types of legal instruments to aid effectiveness, such as the Code of Conduct, as Conventions are difficult to change.
— Examining how the multiple international legal frameworks could work better together.

Panel 9 – Communication, Engagement and Trust Building

The safe operation of nuclear facilities is ensured through the cooperation of multiple stakeholders, including operating organizations, regulatory bodies, and TSOs. Interfacing with decision makers and professional organizations, as well as communication with the public through local communities and the media needs to be transparent and clear. Availability of information from different sources and access to new formats of communication present opportunities but also challenges for conveying accurate and reliable information to decision makers, the media, and the general public during both normal operation and in emergencies.

This panel discussed how information can be shared accurately and in a timely manner in a way that is understandable to the target audience and builds trust with the public. This was in the context of reinforcing effective communication. Depending on who is communicating to whom, when and why, the facts may be expressed differently and hence can induce uncertainty and mistrust.

The panel found that:

— The nuclear industry communicates risks well but is not as good at communicating the benefits of nuclear energy.
— Communication experts and social scientists have a key role to play in positively and effectively engaging the public, as do young enthusiastic influencers.
— Trust takes many years to build, through effective communication and continuous engagement. It has to be established before an accident does occur.
— It is important to ensure an understanding of nuclear phraseology and to realize that there cannot be total consistency in messaging, given that stakeholders and target audiences are diverse. The differences in messaging need to be monitored.
— It is important to find and use the most appropriate communication channels and tools and ensure that the public has a clear idea about the appropriate places to go for information.
— The IAEA is a trusted source of information, especially for newcomers, but independent and government sources also have a role to play.
— Engaging with journalists (local, national, etc.) openly and transparently and providing feedback on their communication is one of the best ways to reach the interested public.

The panel also debated the role of the public in nuclear decision making, its benefits, and risks in building trust but potential short-term perspective.

— The public and the media need to be included as stakeholders in nuclear safety, but decisions are made by those in the authority. Governments and regulators do need to listen to and understand their perspectives.

Learning from other areas was also a theme, especially from the COVID-19 pandemic where concepts of risk and balancing risks were fed, sometimes daily, into the public consciousness.

— There are clear parallels between nuclear risk perception and vaccine hesitancy.
— The pandemic did make clear that the public is not a homogenous entity and people do have a capacity for understanding risk vs benefits.
— It is still unclear how best to bring relevant stakeholders into the decision making process.

Sustaining effective communications in the long term was a particular issue highlighted by the panel, with panellists noting that, to maintain relationships with relevant stakeholders, one needs to communicate frequently, openly and in plain language. Finally, the society need to trust the "messenger" in order to trust the message.

Panel 10 – Facing New Challenges

The rapidly changing world inevitably reflects on the nuclear energy landscape. Considering this changing environment and new technologies, enhancing safety in the next years remains a key task for governments, regulatory bodies, and operating organizations. Natural phenomena, economic trends and social expectations are difficult to foresee. New technologies impact energy demand and supply, and shape life. Events, such as pandemics, impact mobility of people and goods and affect nuclear supply chains.

During this panel, participants discussed trends and global developments that might have a direct impact on nuclear power in its current format:

— Current NPPs are robust against the challenges of climate change but there are other challenges such as cyber safety and security, parts obsolescence in ageing nuclear facilities, and the challenge of capacity building for small and newcomer countries.
— The cost of nuclear energy has increased against other clean energy. Standardization would make nuclear energy more competitively priced and quicken deployment. Regulation like that established for the aviation industry might help, as well as moving away from technologies that have been only proven for nuclear applications.
— SMRs could also be the answer to the increasing costs facing nuclear power development programmes; however, they have no history in operation or regulation.
— As new technologies continue to emerge, the regulation of innovative designs will require global cooperation.
— Social media and new communication channels heavily influence public trust and support. The nuclear industry has to make its case as a safe, reliable and climate friendly source of energy in a clear and easily understood way.

Panel 11 Call for Actions – Maintaining the Momentum

Recognizing that discussion is not enough to defend against complacency and demonstrate the safety of the nuclear industry, this panel proposed actions for moving forward. Panellists used the key messages from the conference coupled with insights from speakers throughout the week to identify where future efforts and momentum would be most beneficial.

The panel discussion focused on four areas: enhancing openness, embedding the lessons of the Fukushima Daiichi NPP accident, better preparing for the wider use of safe nuclear power, and transferring knowledge for the future. The conclusions of the discussions, reflected in this section, are summarized in the President's Call for Action. The discussions are summarized in the statements below.

1) Enhancing openness: This is not just building trust but part of building robust national and international systems, with interactions and interfaces that make the systems work. It is important to talk and listen to make nuclear systems more robust.

— There is a need to be better at getting out the messages around risks and benefits with better guidance on communicating, especially to young people. Part of this is how to better explain and communicate how safe is safe enough.
— There is a need to improve emergency planning and response and its transparency, expanding beyond the immediate stakeholder community to the public health community, as the response has an impact on the non-radiological effects of an emergency.
— A more holistic approach to low-dose radiation is needed to differentiate between attribution and inference of risk. In protecting people against radiation exposure, both the radiation risks

and the psychosocial effects of radiation protection need to be taken into account. The radiation health effects need to be balanced with the psychosocial effects and the effect of protective actions.

— In recovering from a nuclear emergency, the answer to what is safe enough needs to be established in advance in the consultation phase and communicated to the public as a range of acceptable end states.

— It is important to take into account the social aspect of improving safety. Trust building is not convincing people but providing them with information so that they can form their own opinions in an informed way. We need to talk and listen to local people; it is less about challenges and more about understanding and providing information, explaining what is being done and why.

2) Being better preparing for the wider use of nuclear power: are we prepared for this change? It could mean that, instead of around 400 operating NPPs globally, we could have perhaps over 1000 with many more countries having nuclear power programmes.

— There is a need for a review of the global structures and systems to drive forward a culture for safety and to ensure that there are appropriate governmental and institutional frameworks, in accordance with and in adherence to the IAEA safety standards, sufficient availability of technical support (especially for new users), and effective and efficient international collaboration. There is a need for more guidance to help foster a wider culture for safety to regulatory bodies, operatoring organizations, policy and decision makers, and the public. And to learn how to apply and nurture this culture.

— Culture for safety needs to expand also for security.

— There is a need to equip people outside the nuclear industry with guidance and tools that incorporate the wider impact of a nuclear or radiological emergency on the public and on public health systems.

— There is a need to find more effective ways to communicate with society about the wider use of nuclear applications, clarify the scope and target of the communication and make use of knowledge from social and behavioural sciences. This communication has to be honest, open and transparent.

— There is a need to be prepared to recover as well as to respond to accidents. The transition to different phases occurs gradually and in different ways. There is a need for clear standards and guidelines for radionuclides in commodities, with international cooperation and commitment. This might avoid unnecessary harm due to a of lack of understanding of the risk.

— It is important to recognize, anticipate and adapt with agility to changes in technology (e.g. analog to digital) and society, with knowledge management, capacity building and young generation engagement. All of this has to be based on a questioning and listening attitude.

3) Embedding the lessons from the Fukushima Daiichi NPP accident:

— New technology designs need to embed these lessons and regulators need to verify this. A complete picture of these activities will help to build greater confidence in safety, especially the application of defence in depth.

— Much has been done in existing NPPs but not all of these activities have been documented or sufficiently explained to the public.

— It is important to engage the public in a holistic approach to decision making.

— Off-site radiation protection issues and the effects of evacuation and large scale contamination need to be reviewed for good and poor practices. The question needs to be changed from the just a question of radiation doses to what is best for people's well-being overall.

— Risk and risk informed decision making need to be better understood. The non-radiation effects of the Fukushima Daiichi NPP accident have had more impact than radiation exposure on people's health and well-being. There are lessons from other natural disasters about how to talk about risks, but there are also communication challenges of fear and stigma of radiation and misinformation.

— Learning never stops and there is continuous improvement for safety. It is the responsibility of operators, regulators and other stakeholders to improve the process of learning. Knowledge transfer from other industries is helpful as well as training in international cooperation. It is not a question of improving safety at any cost but a balance to further strengthen safety where reasonable to do so. Improving safety does not necessarily mean that nuclear applications were previously unsafe. Safety and learning are continuous processes. There is a need to continue to look for ways to be safer based on the use of techniques, such as a cost–benefit analysis.

4) Transferring knowledge for the future: this knowledge is hard won.

— Experiences need to be shared openly. As an example, continued sharing of the experiences in Japan have been and will continue to be valuable to the international community, for instance with regard to decommissioning. Actions to facilitate the transfer of knowledge for other circumstances are needed.

— There is an audience for knowledge transfer. Actions to transfer knowledge are needed. Japan has now considerable experience on recovering from a nuclear accident, and there is a need to ensure that the next generation has access to that information. It may revolutionize decommissioning of NPPs.

— Decommissioning is ongoing at the Fukushima Daiichi NPP, and sharing information and transparency internationally is paramount for ensuring public trust.

— New technology has been developed for these particular circumstances. This large accident is unique in the very long timeframe for recovery and its circumstances, which provides opportunities to put into practice some specific lessons learned. Therefore, there is a need for a repository of knowledge that is searchable and accessible for people who might need it in future, showing what was done, why and how.

— There is an intergenerational aspect of knowledge transfer. The IAEA has developed tools such as peer reviews and advisory services, and capacity building in its various schools. These are available for future generations. A big strength of the nuclear sector is the ability to work together and there are a lot of opportunities for the harmonization of approaches.

— Greater gender equity and skill diversity in the sector is needed.

— Resources are necessary for knowledge transfer. More experienced countries need to help and support embarking countries in investing in technology and people. We all need to widen our interactions and we need smaller, more issue focused meetings to widen our views.

— Another mechanism for knowledge transfer and collaboration is needed in addition to large meetings and conferences. Bilateral and regional collaboration are to be encouraged.

— There needs to be efficiency in working together. There is a need to prioritize, to focus on what matters the most for nuclear safety. There has to be coordination between international organizations working on the same topics. Some of our existing tools have reached maturity and there might be a need to review our legal instruments; the world is markedly different than when they were established. We need to develop advanced tools for the future.

A final question is how to maintain the momentum. The following were proposed:

— By conducting follow up discussions at regular intervals to optimize the value of the work done for this conference;

— By defining the conference outcomes as action-oriented items and facilitate their implementation in straightforward way;

— By sharing the findings from this conference with other topical events for more detailed consideration.

Summary and Conclusions – A Call for Action Closing Remarks

M. Weightman gave his closing remarks, including a summary of his observations and a President's Call for Action to achieve the goal of "Safe nuclear power for all" as part of providing secure, clean energy for humankind and hence part of the solution to climate change. He noted the enablers for this goal: greater international collaboration, optimum decisions, balance, inclusive leadership, the international legal instruments and earning trust. He outlined the four areas covered in his Call to Action: enhancing openness, embedding the lessons of the Fukushima Daiichi NPP accident, better preparing for the wider use of safe nuclear power, and transferring knowledge for the future, with underlying actions to support them. These are contained in the President's Call for Action. He also called for a new mechanism to ensure that the actions were delivered. He suggested an annual side event at the IAEA General Conference to provide an overview and update on the progress made in relation to the actions, as well as to review the list of actions, and that such an event would be organized jointly with other international institutions and would be open for attendance. This will involve being committed to working together and addressing the actions effectively with vigour.

G. Caruso noted that there were 689 conference participants, 131 speakers and panellists, 216 observers from 68 countries and 7 international organizations, and thanked everyone involved in preparing for and implementing the conference.

R. M. Grossi thanked the President of the Conference for his service to the nuclear community, noting that this conference was the right thing to do; it was not an autopsy of the accident, nor an exercise in regret and regurgitation. The community took stock and made progress in a shift that is embodied in the four areas noted by M. Weightman in his address. He encouraged participants to persevere in the effort to better inform stakeholders in a timely manner and to help the IAEA in its aims for better communication. His final message was to stay tuned for the changing world.

Author: Mike Weightman

President of the Conference

4. DETAILED SUMMARY OF THE SESSIONS

4.1. OPENING SESSION

Ms Naga Munchetty: Moderator
Mr R. M. Grossi: IAEA Director General
Mr T. Hikihara: Ambassador of Japan
Mr M. Weightman: Conference President (United Kingdom)
Ms L. Evrard: IAEA Deputy Director General and Head of the Department of Nuclear Safety and Security, Keynote Speaker
Mr G. Caruso: the IAEA Department of Nuclear Safety and Security, Scientific Secretary
Mr H. Yamana: Japan Nuclear Damage Compensation and Decommissioning Facilitation Corporation, Keynote Speaker

4.1.1. President of the Conference – Welcome Address

Mr M. Weightman welcomed everyone to the conference, provided context for the event and gave an overview of the programme.

A copy of Mr Weightman's speech has been provided as follows:

Director General, honoured guests, delegates it is my great honour and pleasure to be your President for this most important, timely and seminal conference.

Today, the world faces a climate change crisis that threatens humankind's existence as we know it. It is man-made. But it can be solved by human endeavour. Only, however, if we work together and utilize all our skills, knowledge, experience, science and technology and, importantly, combined will power to take action collectively. Nuclear power can be an important part of the solution.

As nuclear leaders our duty is to ensure we contribute all we can to this unique effort. To be able to do so we need to win over the hearts and minds of the people we serve – the public – to enable the world to reap the benefits from the peaceful use of nuclear energy, generating green energy for all. To achive that we need to remain humble. As an ancient saying tells us: "The heavier the bundle of rice, the lower the head must bow." So we need to harness our combined efforts and demonstrate to the sceptical world that we have learned the lessons from the Fukushima Daiichi nuclear accident and build a culture of never being complacent, never stop seeking to learn and improve, never stop being open, and honest to those we serve. It will be a long and never-ending journey

However, over the last ten years, the nuclear community has made a good start on this journey.

Our Japanese colleagues have undertaken the decommissioning of Fukushima Daiichi and recovery of the surrounding area with great dedication, skill and expertise, and with great success. The rest of the world has rallied round to help by sharing knowledge and experience. And, we have benefited from the openness and transparency of our Japanese colleagues to learn lessons. We are fortunate to hear the latest position from our honoured guests.

I witnessed this from the first time when I went to Fukushima Daiichi, leading the IAEA Fact Finding mission in May/June 2011, and observed this ever since from near and far.

I have also observed the many benefits from international organizations collaborating and working together to assist, understand, learn lessons and take action. This is something that should continue. We will hear from some of these organizations later.

We will also hear later in the week about some of the prime areas where lessons have been learned:

— Ensuring safety;
— Preparing and responding to potential nuclear emergencies;
— Protecting people from radiation exposure;
— Recovering from a nuclear emergency.

Special events will highlight important topics for us all to consider, including a youth panel.

This is how we demonstrate our commitment to collaboration, openness and transparency, and that we have learned the lessons.

But that is not enough. We have to look forward, building on the solid foundations of the past, and we have to commit to action.

Hence, later we move onto panels on:

— Safety for nuclear development – a vital prerequisite;
— Inclusive nuclear leadership – without which we have no future;
— International legal instruments – a basis for expanding nuclear development and international cooperation;
— Communication, engagement and trust building – explaining, responding, serving the people.

And, to make it not mere words we must take actions, commit to an action plan to drive forward to a future where peaceful nuclear energy makes its fullest contribution to solving this crisis. We need to do this as individuals and as organizations.

For me this conference is different. It is different in format, it is different in our expectations, it is different in the outcomes, its demands commitment to action. It is a turning point, or, as our Japanese colleagues may say, a 'Setsu', a basis for ensuring we do our utmost to ensure the peaceful use of nuclear energy makes the contribution it should, to solving the climate change crisis facing the world, ensuring clean energy for all.

Director General, honoured guests (speakers), delegates I commend this conference to you.

Welcome and thank you.

4.1.2. IAEA Director General – Summary of the Opening Address

Director General Mr R. M. Grossi's speech has been summarized as follows:

The Director General welcomed the participants, noting nuclear safety always comes first and nuclear energy has to form part of the mix to combat climate change. This conference was meant to demonstrate that we have taken stock of efforts as a global community, past, present and future. Lessons of the Fukushima Daiichi NPP accident have been learned and acted upon. There is no need to rehash history, the Fukushima Report already exists. He reflected on his attendance at the 2021 United Nations Climate Change Conference (COP26) and questions there about the safety of nuclear energy stemming from the Fukushima Daiichi accident. The fact that nuclear energy is being considered as part of the answer to progress on climate change is of great importance. He highlighted the fact that this conference was organized with many other organizations, listed as co-sponsors, because issues transcend the relatively small circle of nuclear experts. This demonstrates the breadth and width of the challenge. He emphasized that we are also looking forward, continuously improving and learning, never complacent, innovating and adapting to meet the challenges of the future. The intention of this conference is to look at all areas, from the safety of nuclear installations to emergency preparedness and response; how it was done in Japan and how it is being carried out now, to an analysis of the legal framework and how best to support nuclear activities. He concluded with the observation that clear and transparent communication is paramount so that governments and the public can trust that nuclear energy is safe and can make informed decisions about its use in a de-carbonized world.

4.1.3. Ambassador of Japan – Summary of the Opening Address

Ambassador Mr T. Hikihara's speech has been summarized as follows:

The Ambassador expressed his appreciation for the conference and the participation of all the invited experts. He noted the substantial improvements in nuclear safety over the past 10 years, and that nuclear energy supplies a considerable part of global energy needs. Japan continues to make nuclear safety their top priority, following and integrating safety standards. There has been substantial improvement in nuclear safety in the past 10 years. He commended the IAEA Action Plan on Nuclear Safety and briefly explained Japan's efforts to date and their plans going forward, including sweeping reforms in nuclear safety so that Fukushima will never happen again, such as new safety regulations, IAEA missions and the establishment of a nuclear management bureau in 2018 to support local governments for easier planning for recovery. Work continues in decommissioning the Fukushima Daiichi site and an agreement was signed in July with the IAEA for treatment of ALPS water from the plant. Japan continues to provide regular status updates to the IAEA; their other plants continue to remain safe, and they are taking actions for the environmental recovery of the area. Japan continues to share information with other States and signatories to the CNS. He reflected on the high level of international cooperation, including Japan's provision of training at the Fukushima Daiichi site for newcomer countries. He expressed his hope that the conference would help both more mature and newcomer countries to increase nuclear safety worldwide.

4.1.4. Scientific Secretary of the Conference – Keynote Address

Mr G. Caruso welcomed everyone to the conference, noting that over 700 people were in attendance either virtually or online. He highlighted the need for open and transparent interactive discussions so that the results of the conference will be instrumental to further strengthening nuclear safety. He reiterated the message of the conference: that nuclear safety has a role to play at all levels and nuclear energy is of vital importance to the future.

A copy of Mr Caruso's speech has been provided as follows:

Ladies and gentlemen,

A warm welcome to all of you and its remarkable to see the high level of attendance at this important conference. The conference consists of sessions and panel discussions that will cover thematic areas of vital relevance to further strengthening nuclear safety.

The discussions need to be open and transparent so that we all as part of the nuclear community also take on board views/thoughts that will be instrumental in enhancing nuclear safety worldwide. The success of this high level event depends on the interactive discussions that will take place.

Of utmost importance, is the critical focus on issues that are still unresolved and lessons that are yet to be learned and to be addressed 10 years after the accident.

We worked very closely with the members of the Special Experts Committee and the Chairs to identify the relevant topics to be discussed in this conference. I would like to highlight that the thematic topics covered in the programme, relate to key nuclear safety aspects. We hope that the message to be carried from here is that robust nuclear safety at the national, regional, and global levels is vital.

The future of nuclear safety depends on all of us coming together and working for a safer future for generations to come.

I wish you all success and thank you once again for your participation.

4.1.5. Deputy Director General and Head of the Department of Nuclear Safety and Security – Keynote Address

Deputy Director General Ms L. Evrard welcomed everyone to the conference, noting the wide expertise and perspectives of the participants. International cooperation is essential, as nuclear safety is a national responsibility, but accidents affect us all globally. She outlined the IAEA's support for Member States and reiterated that we need to continuously identify ways to further improve nuclear safety. This conference would assist in that regard, shaping IAEA activities for the next decade.

A copy of Ms Evrard's speech has been provided as follows:

Ladies and gentlemen,

A very warm welcome to all of you, attending in person in Vienna or virtually, this international conference dedicated to the lessons learned 10 years after the Fukushima Daiichi NPP accident. This 5-day conference will cover a very wide range of topics, from technical topics to more general ones, as well as specific issues such as societal involvement. The conference will address technical topics such as the safety of nuclear installations or the protection of people against radiation exposure. It will address topics that include an important organizational component, in particular preparing and responding to a nuclear emergency. It will also address key issues such as communication, stakeholders' involvement, and trust building. This combination of topics will offer different perspectives, complementary to each other.

Ladies and gentlemen,

Learning from the Fukushima Daiichi NPP accident in a holistic manner requires this diversity of approaches. The wider the perspectives are, the richer the learning is. Minimizing the potential risks of nuclear energy implies working along all these different dimensions because they are closely linked. Sharing various experiences and combining different views is indeed a key asset to improve nuclear safety continuously. This week, we will mobilize decades of cumulative experience from all over the world, with the participation of many nuclear safety experts, policy makers as well as technical experts from the nuclear industry. This cumulative experience from diverse backgrounds is invaluable to build collectively new paths to strengthen nuclear safety worldwide.

Ladies and gentlemen,

I would like to highlight the significance of international cooperation for nuclear safety:

— Nuclear safety is a national responsibility, but nuclear accidents can transcend national borders. In addition, some countries have more extensive experience in operating nuclear facilities than others, some countries have encountered nuclear events or accidents. Therefore, lessons learned from some countries can benefit others, to maintain high levels of safety and security across the world and to improve safety and security worldwide as needed.

— The Fukushima Daiichi accident underlined the vital importance of effective international cooperation.

— Some of the factors that contributed to the Fukushima Daiichi NPP accident were not unique to Japan. And from a broader perspective, questioning attitude and openness to learning from others' experience are cornerstones for safety.

— The IAEA has a unique position for international cooperation on nuclear safety, with its 173 Member States, from all over the world, covering an extremely wide range of nuclear facilities and activities, and a multitude of experiences.

— The IAEA safety standards provide a robust framework of fundamental principles, requirements, and guidance, that constitute a high level of safety, internationally recognized.

— The Agency also offers a wide range of peer reviews and advisory services, in which teams of experts compare existing practices with the requirements and recommendations provided in IAEA safety standards. The conclusions of these missions highlight good practices worth sharing, as well as recommendations and suggestions, that help Member States strengthen and enhance their national infrastructure and practises.

— The legally binding international conventions for safety under the auspices of the IAEA, including the CNS and the Joint Convention on the Safety of Spent Fuel Management and on the Safety of Radioactive Waste Management, are also key instruments to strengthen international cooperation and continuously improve nuclear safety worldwide.

Ladies and gentlemen,

The main purpose of the conference this week is to identify ways to further strengthen global nuclear safety. After the review of a wide range of issues pertaining to the progress achieved in nuclear safety since the Fukushima Daiichi accident, this conference will be an opportunity to better plan for the future with a "Call for Actions" to be announced at the end of the week. The "Call for Actions" will focus on how the international community will continue to ensure and improve the safety of nuclear installations, by building on lessons learned and to continue to move forward to strengthen global safety.

Ladies and gentlemen,

The conference deliverables will contribute to shaping the programme of the IAEA Department of Nuclear Safety and Security within the next decade. I am confident that this conference will help us to see where we stand today regarding nuclear safety and the concrete steps that we need to address in the coming decade. I would like to commend the preparatory work conducted by the Special Experts Committee, that provided valuable inputs thanks to their extensive experience.

I would like to give special thanks to the President of the Conference, the experts of the Committee, Japan for their support, the panellists and the speakers of the conference, and all those who have been involved in the preparatory work

Thank you very much. I wish you all a very successful conference.

4.1.6. President of the Nuclear Damage Compensation and Decommissioning Facilitation Corporation – Summary of the Keynote Address

Mr Yamana's speech has been summarized as follows:

Mr Yamana noted that it had been 10 years and 7 months since the accident at Fukushima Daiichi. He spoke from his experience working with the Japan Nuclear Damage Compensation and Decommissioning Facilitation Corporation on lessons learned and how best to avoid future accidents, based on the status of Fukushima Daiichi post-accident. Institutional issues needed to be addressed, such as the safety culture of operators and the independence of regulatory institutions. Technological issues were examined, in particular defence in depth and the diversity of safety systems. He also reflected on lessons learned in other areas. When considering evacuation issues, more attention needs to be paid to its psychological and social effects. Effective accident management needs to have the collaboration of the government and onsite operators and directors in emergency preparedness and response and the capabilities of staff are key to dealing with an emergency. Social recovery includes compensation for survivors and damage. The government bears responsibility here; in Japan, TEPCO compensated victims. The restoration of the contaminated environment requires a massive decontamination effort; in the case of Fukushima Daiichi, it has been reduced to one-fifth of the original area and people are slowly returning, however, reconstruction will continue for a long time to come. A special reconstruction tax has been established to fund this work. As for decommissioning of the site, treatment of the contaminated area is ongoing, and two-thirds of the spent fuel has been recovered. He noted decommissioning the site continues to be a big challenge, and that stakeholder involvement in the final solution to decommissioning is very important. He highlighted the serious global reputational damage caused by the accident, with public trust still not fully regained and concluded with a call to work together to establish even better and safer systems for nuclear safety moving forward, with lessons learned from Fukushima Daiichi NPP accident.

4.2. SESSION A – CONTRIBUTION OF INTERNATIONAL ORGANIZATIONS TO GLOBAL SAFETY

Mr M. Weightman, Session Chairperson, United Kingdom/Conference President
Mr W. Magwood, Organisation for Economic Co-operation and Development/Nuclear Energy Agency (OECD/NEA)
Mr I. Engkvist, World Association of Nuclear Operators (WANO)
Mr G. Caruso, IAEA/Department of Nuclear Safety and Security
Ms G. Hirth, United Nations Scientific Committee on the Effects of Atomic Radiation (UNSCEAR)
Mr G. Graham, Preparatory Commission for the Comprehensive Nuclear-Test-Ban Treaty Organization (CTBTO)
Mr G. Dercon and Mr C. Blackburn, Food and Agriculture Organization of the United Nations (FAO)
Mr J. Pintado Nunes, International Labour Organization (ILO)
Mr L. P. Riishojgaard, World Meteorological Organization (WMO)
Ms Maria Neira, World Health Organization (WHO)

In the aftermath of the Fukushima Daiichi NPP accident, multiple international organizations contributed to the global response, underscoring their ongoing and important role in supporting nuclear safety. The IAEA adopted its Action Plan on Nuclear Safety, which included reviewing and revising relevant safety standards, strengthening peer review services, and enhancing efforts to assist Member States in building capacity for safety. Other international organizations provided support through the promotion of stress tests, new research and development activities, monitoring of post-event health effects, and enhanced cooperation and exchange of experience and information. These activities and findings were subsequently captured in the reports of international organizations, complimenting the actions by countries to review and enhance their national safety approaches. This session, chaired by the Conference President Mr M. Weightman, provided an opportunity for international organizations who contributed to global efforts during and after the accident to share their work and demonstrate that the peaceful uses of nuclear energy are safer now than ever before.

Mr C. Magwood reflected on the changes that have been implemented worldwide to improve safety as a result of the accident and the role of the NEA, noted challenges and issues for the future, and the importance of nuclear power role in responding to climate change.

Mr Magwood's speech has been summarized as follows:

Mr Magwood reflected on when he first heard of the accident and the progress made since it occurred. Daily interactions over the last 10 years have completely reshaped the NEA's relationship with Japan. The NEA issued a report shortly after the accident stating that NPPs are safe but need to be made more resilient to unknown disasters. He noted the enormous importance of the human aspect of safety and safety culture. Five years ago, the NEA reported on changes implemented globally as a result of the Fukushima Daiichi NPP accident. Training, equipment and procedures have all changed in NPPs. He highlighted that "the ability to recover is as important as avoidance" and "safety culture is as important as technical expertise". In March 2021, the NEA issued a report looking forward. It was made accessible to everyone, written not only for technical experts but for the general public as well. It recognized the improvement in regulatory authorities, the importance of non-radiological health aspects, and seizing the opportunity for economic development. He noted that in the future, the removal of the damaged core will be a huge challenge and the help of the international community is needed. Public and stakeholder need to be engaged. The public is part of the discussion. He concluded by saying that nuclear power is essential to the future of the environment, economy and civilization. It has to play a strong role in climate change but that cannot happen if the public does not believe it is safe. It is not only about what happened 10 years ago, but about the future use of advanced technologies like SMRs and others to help mitigate the crisis and improve lives globally.

Mr I. Engkvist of WANO joined virtually and noted that operators are responsible for safety, but regulators, governments and organizations all contribute to safety and all need to work together.

Mr G. Caruso provided the IAEA perspective on the accident and noted the actions taken to improve safety as a result. Mr Caruso's speech has been summarized as follows:

Mr Caruso provided the IAEA perspective on the accident, with an overview of the actions taken at the time and over the past 10 years. He noted the development and implementation of the IAEA Action Plan on Nuclear Safety; numerous international meetings and documents on analysis and lessons learned from the accident, concurrent with IAEA support to Member States to enhance nuclear safety in peer reviews and advisory missions; and the review of the IAEA safety standards to include all the lessons from the Fukushima Daiichi NPP accident. Going forward, the IAEA has also been contracted to oversee the ALPS treated water discharge into the sea and continues to provide support to the Fukushima Prefecture.

Ms Hirth gave an overview of the role and the work of UNSCEAR regarding the Fukushima Daiichi NPP accident and highlighted the findings in the Committee's 2020 scientific report.

A copy of Ms Hirth's speech has been provided as follows:

President, Chairs of the Conference, Excellencies, distinguished experts and colleagues, it is a great honour to be here and to represent the United Nations Scientific Committee on the Effects of Atomic Radiation (UNSCEAR), and on behalf of the Scientific Committee, I would like to congratulate the International Atomic Energy Agency (IAEA) for hosting this conference and bringing us all together this week; enabling us all to share our learnings and experiences from the past decade and to discuss how we can all contribute to strengthening the framework for radiation safety by continuing to build on the solid foundations of our scientific understanding of radiation and its effects on people and the environment.

To begin, I think it is important to recall the mandate of UNSCEAR. UNSCEAR is a unique Scientific Committee of the United Nations, established by the General Assembly 65 years ago. It is the highest authority in the United Nations on radiation levels, health effects and risks, and has the mandate to independently assess sources of exposure to radiation, and their effects on health and the environment. The Committee's scientific reports provide the foundation on which the international framework for radiation safety is built. It has an important role in disseminating its independent findings based on the latest peer reviewed science and evaluations to the United Nations General Assembly, Member States, the scientific community, and the public.

The Scientific Committee completed its first report on the Fukushima accident and reported its findings to the United Nations General Assembly in 2013; the UNSCEAR 2013 Report was published in April 2014 in both English and Japanese.

In 2014, the Scientific Committee initiated a follow-up process, establishing an expert group who collected, screened and systematically reviewed new peer reviewed scientific literature, research reports and available monitoring data across several thematic areas, to identify whether this new information challenged the assumptions of the UNSCEAR 2013 Report, materially affected its conclusions or if it addressed research needs identified in the 2013 Report.

Covering the scientific literature from October 2012 through December 2016, the group reviewed more than 300 publications, and the results of this follow up work were published in the Committee's 2015, 2016 and 2017 White Papers. In each of these White Papers, the UNSCEAR 2013 Report conclusions were recalled, and then the new information was reviewed against that. This helped the Committee understand the relevance and importance of the new material, and again highlighted areas for future research.

These White Papers confirmed that, on the whole, the overall picture remained broadly unchanged but there were also some interesting and relevant scientific developments and new information that could be taken into account in any future studies to refine and improve our understanding of the levels and effects of the radiation exposures from this accident, and in 2018, the Scientific Committee initiated a new project on evaluation of the consequences from the Fukushima accident with the aim to summarize the current understanding based on information up to end 2019 and to consider the implications for the findings of the UNSCEAR 2013 Report. The expert group of over 30 internationally recognized experts with supporting task groups also validated and revised the estimates of doses to the public (including variability and uncertainty) and their health implications.

This project was successfully completed in 2020, after critical review by 13 international experts, and following approval by the Scientific Committee, the UNSCEAR 2020 Report on Fukushima [2] was published in English in March 2021.

In the UNSCEAR 2020 Report, the Scientific Committee revised and updated its previous estimates and their associated uncertainties, providing a more realistic and robust assessment of the exposure of the public resulting from the Fukushima Daiichi NPP accident when compared with the UNSCEAR 2013 Report. While overall, the total doses remained low and were broadly comparable with those in the UNSCEAR 2013 Report, there were some large differences in the component parts (e.g. pathway, location, time).

These differences and improvements in the estimates of public exposure arose because the Scientific Committee had much more information to consider. More extensive measurements of the radionuclides in the environment, and more measurements of radionuclides in people. This included data from extensive personal dosimetry campaigns in a number of municipalities to measure external doses for people (specifically from Minamisoma city and Naraha town); and published assessments by Japanese and other researchers, of doses to people from one or other exposure pathway.

---

[2] UNSCEAR 2020/2021 Report Volume II: "Sourses, Effects and Risks of Ionizing Radiation" Annex B: Levels and effects of radiation exposure due to the accident at the Fukushima Daiichi Nuclear Power Station: implications of information published since the UNSCEAR 2013 Report

The differences also related to improved, more robust and realistic models that were used based on an improved source term with better information on the patterns of releases and movement in the environment, which was used together with improved atmospheric transport dispersion modelling, to estimate the concentrations of radionuclides in air. Models were also improved to estimate external doses, which took into account Japanese specific information that had become available, e.g. (i) more information about soil types, climate, (ii) the use of Japan-specific dose coefficients for the intake of radioiodine, and (iii) specific behaviour of the Japanese people, including information on food and dietary habits and expanded information on evacuation scenarios.

The Scientific Committee also noted that the extensive 5-year programme of decontamination had reduced annual doses to less than 1 mSv in inhabited areas; and enabled return to many evacuated municipalities, and in relation to the environment, there was broad consensus on the levels of exposure of non-human biota, regional impacts on wildlife populations were unlikely but detrimental effects on individual organisms have been observed and others may have occurred.

Overall, with regard to the health implications to the general population of Fukushima prefecture from radiation exposures – the findings of the UNSCEAR 2020 Report are generally consistent with those in the 2013 Report – but there is now more information available to support the Committee's conclusions, that:

— The accident led to no adverse documented public health effects that were directly attributable to radiation exposure from the accident;
— Future cancer rates that could be inferred from radiation exposure from this accident are unlikely to be discernible;
— Increased incidence of thyroid cancer observed in children in Japan was judged to be the result of extensive ultra-sensitive screening.

As part of the process of putting preparing the UNSCEAR 2020 Report, the 2013 Report and the White Papers in between, the Scientific Committee has noted many other important issues that are not part of its mandate – in relation to the broader issues of the health, well-being and social impacts of the nuclear accident. These are all concerns, and the Committee is aware of them – but they are the remit of other organizations.

The UNSCEAR 2020 Report is an authoritative, independent and up to date assessment of the levels and effects of radiation exposure due to the Fukushima accident, based on latest science and monitoring data to end 2019. The main findings are robust and unlikely to change significantly in the foreseeable future.

With respect to key lessons and what factors are critical in enabling the Scientific Committee to realistically estimate doses to exposed persons for the purpose of evaluating health risks to people and the environment.

The Scientific Committee stresses the importance of gathering quality measurement data, especially in the early days of an accident, measurement of radiation in people and the environment, that is taken as soon as possible during and after an accident – in order to make as realistic as possible dose estimates. In our UNSCEAR 2013 Report, there was a higher reliance on modelling data for certain population groups, and also certain exposure pathways, and we know that this tends to make dose estimates more conservative.

It is also important to understand the baseline rates of a cancer, for example thyroid cancer, in a population and understand the sensitivity of screening techniques that may be applied to follow up and detect potential cancers in an exposed population.

In conclusion, I would like to acknowledge the efforts of all of the experts who have contributed to the Scientific Committee's work over the past decade and the broader scientific community and researchers, in Japan and internationally for their efforts and research publications, and finally, I would like to remember those citizens of Japan who lost their lives as a consequence of the earthquake and tsunami and acknowledge those who survived but whose lives have been forever impacted by these events.

Thank you very much for your kind attention.

Mr G. Graham provided an overview of the CBTBO's activities since the Fukushima Daiichi accident.

A copy of the summary of Mr Graham's speech has been provided as follows:

CTBTO radionuclide monitoring segment – overview on 10 years of achievements

The main purpose of the CTBTO's International Monitoring System (IMS) is for nuclear explosion monitoring. The IMS data and International Data Centre (IDC) products are made available to authorized users from National Data Centres for the purposes of CTBT verification.

In addition, this unique asset of global data may also be used for other purposes as decided by the Preparatory Commission for the CTBTO.

The IMS radionuclide stations in the whole northern hemisphere detected radioactive isotopes and noble gases stemming from the Fukushima Daiichi NPP accident, among them Iodine-131 and Caesium-137.

The IDC started sharing its monitoring data and analysis reports with the IAEA and WHO.

In March 2012, the CTBTO became a member of the Inter-Agency Committee on Radiological and Nuclear Emergencies (IACRNE). Further civil applications for disaster risk reduction have been proposed.

According to the Joint Radiation Emergency Management Plan of the International Organizations, the critical response tasks of the CTBTO during an emergency phase is to provide real-time particulate and noble gas monitoring data including confirmation of no detection.

The IMS data and IDC products are also accessible to the scientific community through special agreements.

The presentation will compile key achievements in the CTBTO radionuclide monitoring technology over the last decade, in terms of additional station certification and novel software tools for data analysis and dissemination.

Mr G. Dercon and Mr C. Blackburn outlined activities of the FAO stemming from lessons learned from the Fukushima Daiichi accident.

A copy of the summary of Mr Dercon and Mr Blackburn's presentation has been provided as follows:

This presentation will outline three activities that stem from lessons learned:

— Improved international standards and guidance for radionuclides in food and agriculture;
— Targeted technical aids for use by agricultural departments in general;

— Coordinated international research activities to improve and extend remedial options for agricultural land in less well studied agricultural domains.

In the event of a nuclear or radiological accident and under the Joint Radiation Emergency Management Plan of the International Organizations, the Food and Agriculture Organization (FAO) of the United Nations assigns staff to the IAEA Incident and Emergency Centre at the IAEA Headquarters in Vienna. The FAO, through its Joint FAO/IAEA Centre, ensures implementation, coordination and the dissemination of information with the IAEA on nuclear issues related to food and agriculture.

Post Fukushima Daiichi, from March 2011 onwards, the FAO worked in collaboration with its Member States, the IAEA and other international organizations. In the months and years that followed, standards and guidance related to radionuclides in food received considerable scrutiny. A set of comprehensive standards and norms are available for emergency exposure situations. However, the improvements that were identified related to a need to promote a greater understanding of and development of international guidance for both natural and human-induced radioactivity in foods in non-emergency (i.e. existing exposure) situations. In addition, the Joint FAO/IAEA Centre determined that it could do more to support agricultural departments in Member States. for example, by helping to develop technology to aid the collection, dissemination, and visualization of food monitoring data and to support decision making on agricultural controls. Further, international technical workshops were convened in October 2016 and 2021 to promote and share knowledge and experience related to remediation of radioactive contamination in agriculture. Subsequently, the Joint FAO/IAEA Centre also commissioned an international research project to combine experimental studies with field monitoring and modelling. This research aims to better understand and predict the role of environmental conditions on radiocaesium and radiostrontium transfer in food chains and their dynamics. One of the research objectives is to customize the remedial options in agriculture to suit less well studied agricultural production systems, and to adapt and develop innovative decision support systems for optimizing remediation of agricultural lands, based on machine learning and operations research techniques.

Mr J. Pintado Nunes of the ILO joined virtually and gave an overview of the ILO framework, with specifics on provisions for radiation protection, noting that there is a growing interest in occupational safety and health.

Mr L. Riishojgaard of the WMO joined virtually and gave an overview on the history of the WMO and its role before and after the accident.

Ms M. Neira of the WHO joined virtually and stressed the importance of a coordinated emergency response in the international context and an understanding of risk assessment and its communication. She also noted the transformative lessons of the non-radiological and mental health impacts following the accident.

The Chair concluded the session with some open questions for the audience: Is there a need to provide an up-to-date summary of all this work? Could this be the basis for enhanced international cooperation?

4.3. SESSION B – ENSURING THE SAFETY OF NUCLEAR INSTALLATIONS

Ms R. Velshi, Session Chairperson, Canada/Canadian Nuclear Safety Commission
Mr K. Watanabe, Japan/Nuclear Regulation Authority (NRA)
Mr F. Aparkin, Russian Federation/ROSATOM State Atomic Energy Corporation
Mr P. Tippana, Finland/Radiation and Nuclear Safety Authority (STUK)
Ms A. Pelle, France/Électricité de France (EDF)

The Fukushima Daiichi NPP accident emphasized the importance of continuously challenging the existing assumptions regarding nuclear safety to prevent future accidents. The attention of the nuclear industry refocused on safety improvements in the existing nuclear installations and enhanced safety considerations for new projects and designs. Specifically, countries reviewed and reinforced, as necessary, nuclear installations' capability to withstand or control possible accidents originating from extreme conditions and external events to minimize risk. In support of these activities, the IAEA and the Member States reviewed and revised their safety frameworks, including updates to the safety standards, to ensure that the lessons identified are incorporated at the national level to ensure global nuclear safety. During the session, speakers summarized the lessons from the accident and associated actions and highlighted the different approaches and decisions that were taken to enhance nuclear safety and face emerging challenges.

The Chair opened the session by stating that the accident caused a major paradigm shift in the nuclear industry; instead of preparing for every conceivable accident, we instead prepare for all accidents regardless of how they start. To this end, we have obtained portable generators, pumps, and other various pieces of equipment to assist us in mitigating accidents if and when they occur. At what point do we consider our reactors to be safe-enough and accept that accidents may occur? She further requested that the audience ask questions relating to 'what else can be done to ensure safety?'

Mr K. Watanabe outlined the reform of Japan's nuclear regulator to provide independence with integrated functionality. He stressed the importance of openness and transparency for a regulator. He also outlined reforms to Japan's nuclear regulation which included new regulation on severe accidents, with all regulation being based on state-of-the-art information.

Mr F. Aparkin stated that safety assessment found Russian NPPs met regulatory safety standards and further safety measures were introduced for beyond design basis accidents. He noted that Russian Federation is building a strengthened safety concept for future NPPs, including SMRs.

Mr P. Tippana outlined how the nuclear regulator's organizational culture impacts on nuclear safety and stressed that our national cultures influence such things as how we see safety, make decisions, communicate, etc. He explained that STUK conducted an exercise to understand the Finnish national culture vis-à-vis safety culture. He posed a question to the audience: Is the current way we regulate the best way or is there another?

Ms A. Pelle gave an overview of what EDF has done to improve the safety of its NPPs, including equipment upgrades and improvements in their organization to include reactive and proactive elements. She also outlined further improvements to be made as part of extended service life.

A variety of questions were posed by the Chair to the session's speakers during the Q&A period that followed.

The first was to Mr Watanabe, asking if Japan had seen a change in the regulatory culture following the reforms and how would that be assessed. He responded that there was evidence of more openness and transparency in meetings with the public and a more questioning attitude among staff. He noted that they do not conduct public opinion surveys.

The next question was to Mr Aparkin on how mature nuclear countries could better support embarking countries. He responded that they could assist in founding the regulatory body and operating organizations, all based on the safety fundamentals of the IAEA.

Mr Tippana was then asked what insights were gleaned from STUK's national culture assessment and how would they change as a result. He responded that it found that everyone felt very busy and that they did not have the tools to prioritize their work based on safety significance. STUK is working on tools to be risk informed to address this issue.

Ms Pelle was asked at which point did EDF say "this is safe enough" and how did they balance safety with operation. The response was that they used a deterministic and probabilistic approach.

Questions from the floor included one to Mr Tippana asking if he thought ingrained culture could be changed or if more diversity is needed. His response was that cultures change very slowly, but it is necessary to understand them first. A second to all speakers asked if different strategies were needed for large and small reactors. Mr Aparkin noted that conceptually there is not much of a difference between the two, the overall safety approach of defence in depth remains the same.

The Chair concluded the session by stating that complacency is not an option; a strong safety culture with a focus on openness and transparency is vital, decision making needs to focus on safety and safety needs to be considered from design to decommissioning.

4.4. SESSION C – PREPARING AND RESPONDING TO A POTENTIAL NUCLEAR EMERGENCY

Mr C. Hanson Session Chairperson: United States of America/Nuclear Regulatory Commission
Mr T. Makino: Japan/Nuclear Disaster Management Bureau
Ms C. D. Sjogren: Sweden/Swedish Radiation Safety Authority
Ms H. Almarzooqi: United Arab Emirates/Federal Authority for Nuclear Regulation
Mr M. Grzechnik: Australia/Australian Radiation Protection and Nuclear Safety Agency
Mr T. Zodiates: International Labour Organization

An integrated and coordinated emergency management system for preparedness and response for a nuclear emergency should be in place at the national level. The Fukushima Daiichi NPP accident has shown that these arrangements should cover the case of responding simultaneously to a nuclear emergency and a natural disaster. In a nuclear emergency, protective actions have to be implemented in an effective and timely manner and do more good than harm. As these actions can be extremely disruptive for normal life, a comprehensive approach to decision making has to be followed to ensure balance between potential radiological consequences, non-radiological consequences, and health hazards, with special consideration to sensitive population groups. IAEA requirements and generic criteria address the termination of a nuclear emergency and the subsequent transition to an existing exposure situation; however, the Fukushima Daiichi NPP accident highlighted that further guidance is needed. This session, chaired by Mr C. Hanson, shared experiences on various aspects of managing emergency response, including protecting emergency workers and helpers and justifying protective actions.

The Chair stated that it is our responsibility to make sure that emergency preparedness and response (EPR) does more good than harm and that non-radiological impacts can be just as harmful. He noted that the session would cover how to weigh the risks associated with emergencies, planning, considerations associated with first responders, and the use of an all-hazards approach. He mentioned that planning should include provisions for simultaneous emergencies and how to deal with the various stages of an emergency, including transitioning to the end state.

Mr T. Makino provided key lessons learned from the accident in Japan, an overview of the current Japanese emergency preparedness and response (EPR) framework, and updates in response to COVID-19. He emphasized that it is important to plan as much as possible but realize that the plan cannot be perfect.

Ms C. D. Sjogren shared the Swedish strategy, based on an in-depth analysis done in 2014, to create a great deal of trust between the authority and the public and decision makers so as to be able to take practical actions.

Ms H. Almarzooqi provided an overview of actions taken in the United Arab Emirates (UAE) following the accident and an overview of ConvEx-3, which was hosted by the UAE in October 2021, noting the importance of international exercises like ConvEx-3 to building capacity at the international level.

Mr T. Zodiates provided an overview of EPR from the worker's perspective, noting that all decisions should be optimized and justified when it comes to worker safety.

Mr M. Grzechnik, the current Chair of EPReSC, gave an overview of the history of EPReSC and future plans, and discussed a holistic approach to emergency planning. He noted that EPReSC was formed as a direct result of the Fukushima Daiichi accident and that the roadmap for the future is driven by Member States' needs and the analysis of gaps and mapping, including protective strategies for planned, emergency and existing exposure situations. Governments have suggested that this guidance should be elevated to safety standard status in order to be fully utilized.

In response to questions from the floor and online, the following ideas were suggested: Given the difficulty of updating or changing conventions, a different approach would be to develop something in parallel to support it, such as a protocol. To counter anti-nuclear arguments about the expense of nuclear an all-hazards approach can be used, such that this is just another layer to protecting the environment and augmenting EPR as a whole, using economies of scale. COVID-19 has shown that the public can be educated on emergencies, so if we look at dose from their perspective and communicate accordingly, practicing with them and providing familiarity through continuous effort, such as using schools to increase public awareness, we can gain acceptance. To deal with the non-nuclear/radiation consequences of nuclear emergencies, it was suggested that the IAEA could work with other international organizations and include the relevant experts on those issues, e.g. psychosocial, to gather as much information and guidance as possible and see what can be learned. A detailed common geographical information system map for EPR across a whole government would be good in theory but a challenge to implement on various levels.

The Chair summarized the session findings as follows: Communication and public trust are vital, EPR has to be more than the plan (i.e. data integration, exercises and training); planning should be integrated into an EPR framework that considers all-hazards, and consideration needs to be given to the psychosocial effects of an emergency, such as the repatriation of displaced individuals and the mental health impacts of evacuation or sheltering in place.

4.5. SESSION D – PROTECTING PEOPLE AGAINST RADIATION EXPOSURE

Mr N. Ban Session Chairperson: Japan/Nuclear Regulation Authority
Mr A. Gonzalez: Argentina/Argentine Regulatory Authority
Mr E. Metlyaev: Russian Federation/The Federal Medical and Biological Agency
Mr T. Smith: United States of America/Nuclear Regulatory Commission
Mr G. Thomas: United Kingdom/Imperial College London
Ms Z. Carr: World Health Organization
Ms J. Garnier-Laplace: Organisation for Economic Co-operation and Development/Nuclear Energy Agency

An important lesson from the Fukushima Daiichi NPP accident refers to the difficulty non-specialists have in understanding the international system of radiation safety, including the principles and criteria for radiation protection. It is important to communicate the rationale behind the judgement as to whether and how radiation doses to the public should be averted and to make clear that justification of protective measures and actions is based not solely on science but on consideration of the overall benefits and detriments to society and the individual. Furthermore, guidance on monitoring doses to the public in the aftermath of an accident can be limited and this potential lack of information might create public anxiety. The pubic is particularly concerned about the protection of children and pregnant women after a nuclear accident. This session discussed the challenges and successful approaches for protecting the public against radiation exposure and for ensuring timely and effective communication to the public.

The Chair noted that the UNSCEAR report clearly stated that there were no radiation health effects from the accident, but people remain concerned. Therefore, we need to keep in mind people's total well-being in thinking about health effects.

Mr A. Gonzalez provided an overview of the UNSCEAR report, i.e. there were no radiation health effects but other serious health effects, stating that we should be careful in attributing health effects to radiation when there are none or they can only be conjectured.

Mr T. Smith noted that USNRC has done a great deal of research and development to further develop EPR tools. This research can be used to increase public confidence as data can be a factual answer to the question "Is it safe?" Research has shown that evacuations cause more harm than benefit across all types of emergencies, indicating the importance of an all-hazards approach.

Mr G. Thomas noted that scientific communication is often not understood due to the language and jargon used. The audience needs to trust the person/organization providing the information. Data would show that it is very difficult to quantify risk associated with low doses of radiation; however, in relation to many other common safety risks, it is insubstantial.

Ms Z. Carr noted that mental health and psychosocial support during radiation emergencies has become a serious issue, especially in light of the COVID-19 pandemic. Information management during a pandemic may take as many resources as managing the pandemic itself. Protecting people should follow ethical principles of equality to ensure all are protected equally.

Ms J. Garnier-Laplace outlined the NEA work on mental health and psychosocial support impacts and stated that they need to be better considered during and after an emergency, noting that they cannot be dealt with using a 'one-size-fits-all' approach.

There were two questions from the floor during the Q&A session that followed. The first asked how a regulator deals with mental health. The response was that we have to use a risk informed approach such that decision makers can manage their areas of responsibility as they see fit and regulators should look at risk in a broader context than just radiation. The second asked how to deal with media misinformation. The advice was to reach out proactively to provide them with the correct information.

In summary, the Chair noted that, while radiation effects might not occur during a nuclear accident, public fear is enough to cause harm. To that end, we have to be ready to communicate effectively with those affected to assuage fears.

4.6. SESSION E – RECOVERING FROM A NUCLEAR EMERGENCY

Mr C.-M. Larsson, Session Chairperson: Australia/Australian Radiation Protection and Nuclear Safety Agency
Mr A. Ono: Japan/Tokyo Electric Power Company
Mr K. Yumoto: Japan/Ministry of Economy, Trade and Industry
Mr T. Sagawa: Japan/Ministry of Environment
Mr M. Tsubokura: Japan/Fukushima Medical University

Mr O. Novikov: Ukraine/Special State Enterprise Chornobyl Nuclear Power Plant
Ms A. Canoba: Argentina/Argentine Regulatory Authority

The nuclear industry exists in a broader context; therefore, recovering from a nuclear emergency is a complex process and requires coordination among a wide range of stakeholders and consideration of technical and societal aspects. This session discussed topics including the role of technology and innovation, the involvement of the public in remediation efforts, and the identification of challenges that can inform future planning. Speakers shared their experiences from the perspective of international organizations, government authorities, and local and regional leaders involved in the recovery operations from the Fukushima Daiichi NPP accident. All participants, except the Chair and Mr O. Novikov, joined virtually.

In opening the session, the Chair noted that planning for recovery is part of assuring overall safety and emergency preparedness. The more prepared we are, the safer we are.

Mr A. Ono provided an overview of the accident and recovery efforts over the last 10 years. Recovery is an ongoing effort, and they are just beginning, the challenges include many unknown factors. It is an unprecedented mission, and the end state is yet to be decided. Future work includes developing tools to facilitate project management, human resource development, research and development activities, further international cooperation and stakeholder engagement.

Mr K. Yumoto noted that decommissioning is expected to continue for 30–40 more years, with spent fuel removal complete and fuel debris removal to begin soon. Major challenges include water management, fuel removal and fuel debris retrieval, and waste management. He stressed the importance of stakeholder involvement from the local to the international level. He also provided an overview of the coexistence of reconstruction and decommissioning with the recovery of economic centres and local businesses, the return of displaced individuals and reconstruction of local infrastructure.

Mr T. Sagawa provided an overview of the legal framework and decontamination policy, decontamination work methods, and outcomes to date. He noted that 30% of evacuees have returned and 38% of farmland is now being used again. He provided an overview of a demonstration project for soil recycling and stressed the importance of international cooperation to this work.

Mr M. Tsubokura provided an overview of the wide variety of medical issues associated with people in contaminated areas, including those evacuated. Risks are of particular note amongst vulnerable populations, e.g. the elderly. He explained the results of the Fukushima Health Management Survey, which included the incidence of depression and/or anxiety, issues with diabetes, as well as issues with separation from family and reductions in access to health care facilities, which led to a reduction in positive health outcomes. He also mentioned that stigmatization from the media led to negative health impacts.

Mr O. Novikov provided an overview of the Chernobyl accident and the recovery from it, noting lessons learned. He reflected that Chernobyl changed our minds about nuclear and that though the two accidents were different, they were also similar. He highlighted that there is no international standard for the end state of recovery efforts.

Ms A. Canoba provided an overview of the issues and standards relating to contamination in consumer goods, including food and drinking water. She stated that there are differences in dose criteria, terminology, etc. for various goods, which may create confusion as to what is safe. Further work is needed internationally to clarify and harmonize criteria and provide clear guidance for special cases and communication to the public.

Questions from the floor were related to health effects from radiation arising from the accident; it was agreed there were no negative health outcomes from radiation, but rather from fear, anxiety, and other medical and societal issues.

5. DETAILED SUMMARY OF THE PANEL DISCUSSIONS

5.1. PANEL 1 – ENSURING THE SAFETY OF NUCLEAR INSTALLATIONS – MINIMIZING THE POSSIBILITY OF SERIOUS OFF-SITE RADIOACTIVE RELEASES

Ms N. Munchetty: Moderator
Mr M. Foy: United Kingdom/Office for Nuclear Regulation
Mr M. Franovich: United States of America/Nuclear Regulatory Commission
Mr J. Lee: Republic of Korea/Korea Institute of Nuclear Safety
Mr J. C. Niel: France/Institute of Nuclear Safety
Ms R. Sardella: Switzerland/Swiss Federal Nuclear Safety Inspectorate

Findings from the Fukushima Daiichi NPP accident have shown that a drive towards continuous safety improvement leaves no place for complacency. Technical, organizational and regulatory measures are taken to enhance safety, further reducing the likelihood of occurrence of accidents with significant radioactive releases. On the other hand, as such an accident might still occur, the nuclear industry and regulatory bodies need to be prepared for the unexpected.

This panel, moderated by Ms N. Munchetty, discussed measures that can be taken to ensure that serious accidents are very unlikely and highlighted actions to ensure that serious off-site releases will be avoided or minimized, in line with the principles of the Vienna Declaration on Nuclear Safety. The panel also investigated the potential of advanced reactor technologies to practically eliminate the risk of off-site releases and discussed where a questioning attitude contributes to safety and where strict implementation is a prerequisite for the safe operation of nuclear installations.

Mr J. C. Niel reflected that the industry has to take the time to consider key factors to learn from accidents and improve safety. Data tools should be used to further refine analysis and defence in depth should be practiced throughout. Psychosocial impacts cannot be ignored.

Ms R. Sardella noted that the accident affected the whole nuclear industry. Everyone took the time to learn from the accident and this is how we will improve. The industry has made improvements and needs to continue to ask the hard questions and determine if it is enough.

Mr J. Lee joined virtually and gave an overview of Korea's response actions to the accident.

Mr M. Foy noted that, while the industry integrated the lessons learned, everyone should also take the time to look forward and determine how we will continue to improve into the future. The nuclear industry needs to remain vigilant to earn public trust, prevent any major accidents, as well as work on climate change and new technologies.

Mr. M. Franovich emphasized that leadership, risk management and knowledge management and knowledge transfer to the next generation are all important issues for both current and new technologies. A bias for action is necessary to avoid 'paralysis by analysis'.

The panel began their discussion by noting some of the most important changes resulting from the accident, including: an updated safety analysis of hazards; a change in regulatory mindset; the examination of 'black swan' events and giving operators the resources to deal with them; the realization that what was considered impossible was possible and it is necessary now to be open minded and imagine new situations. In discussing international cooperation, panellists noted comfort with the good existing framework but there is a need for more transparency and challenge in the future; historically, openness exists but we need to confirm that the information created is being used properly and the right lessons are learned.

Looking to the future, they also emphasized that it cannot be assumed that others know what we know and the dialogue between regulators and operators needs to be sustained; as generations change, the knowledge needs to be passed on. International dialogue has always existed but to move forward, detailed technical analysis is also needed; a strong safety culture is needed throughout the nuclear industry and needs to be carried forward through the next generation. The panellists discussed the concept of 'safe enough', noting that the regulator reviews the operator and, based on the information available, makes a judgement if it is safe enough. This decision is reviewed on a regular basis and can be updated. However, it is not just about "safe enough". The industry needs to continue to push safety forward and encourage the tolerance of safety risk by the public, communicating what is being done by operators to be safe, how that is being regulated and monitored and what is being learned to improve safety even more going forward. However, one of the challenges to safety is the need to consider a multitude of perspectives in making decisions; those of regulators, NGOs, government, boards, etc. The exciting potential of new technology and SMRs was discussed, noting the passive systems within SMRs and that current designers have designed innovative new reactors to deal with the issues facing existing ones. Concluding comments emphasized that we can never be complacent, there is always room for improvement. A clear allocation of responsibilities is needed, with a questioning attitude and continual review to prevent further accidents and a strong regulator to engage in a two-way dialogue with the public to develop trust.

5.2. PANEL 2 – PREPARING AND RESPONDING TO A POTENTIAL NUCLEAR EMERGENCY – ROBUST PREPAREDNESS ARRANGEMENTS

Ms N. Munchetty: Panel Moderator
Mr K. Kumar De: India/Nuclear Power Corporation of India
Mr T. Makino: Japan/Nuclear Disaster Management Bureau
Ms S. Perkins: United States of America/Nuclear Energy Institute
Ms P. Wieland: Brazil/Brazilian Association for the Nuclear Activities Development

Robust preparedness arrangements need to be in place for responding to an emergency at a NPP that might occur simultaneously with a natural disaster. The response to a nuclear emergency involves many national organizations, as well as international organizations and, therefore, has to be coordinated and effective. This panel discussed the importance of infrastructural elements for emergency preparedness and response, including regulatory requirements; clearly defined roles and responsibilities; pre-established plans and procedures; tools, equipment and facilities; training, drills and exercises; and a management system.

Mr K. Kumar De joined virtually. He noted that, after the Fukushima Daiichi NPP accident, India reviewed all currently operating NPPs and those under construction. Improvements were made in many areas, including monitoring systems, management and EPR planning. Dedicated communication plans were developed for EPR. Field exercises were conducted and reviewed, with updates made to EPR manuals and accident management systems, incorporating lessons learned and further consideration and understanding of IAEA safety standards and international criteria.

Ms P. Wieland noted that the public is in a different state of mind during an actual emergency. Effective communications are crucial in an emergency, and we need to consider the fear and stress involved and tailor communications accordingly. How do we define the severity of an accident? We have reference levels, but how do we communicate them? Words like 'risk', 'contamination', etc. mean different things to different people.

Ms S. Perkins provided an overview of the development of an optimal strategy for loss of power and catastrophic events in response to the accident, with a nuclear response framework for national response to incidents at any reactor. The industry-wide response plan gives new tools to operators for responding to an emergency, including common equipment, governmental coordination and Memoranda of Understanding.

Mr T. Makino discussed how to communicate with workers during and after an emergency. He shared examples of how different modes of communication with the workers at Fukushima Daiichi NPP and the local hospitals affected the uptake of individual health care follow-up and patient care.

In response to various questions posed by Ms Munchetty, panellists noted that nuclear communicates well with other government departments when the all-hazards approach is integrated into national arrangements. Responses to questions from other industries are quite transparent as the nuclear industry is proud of its achievements. Trust is being built with local public safety and elected officials as effective communication relies on local community ties. In Japan, trust is being rebuilt with local communities though it is challenging. Local governments are highly important in Japan, as they implement national strategies and communicate with the local residents; however, the national government now takes control vs local responsibility to protect the people. In Japan, the relationship with the media has become more complicated due to the challenge in communicating scientific concepts and the politicization of information. It is also difficult to communicate that the government is not perfect, but that safety has improved. In the USA, conversations around nuclear are becoming more positive, with the evolution of climate change and de-carbonized energy.

Panellists noted that, during the emergency preparedness phase, it is difficult to get media interest in the industry. There are questions about the use of influencers and virtual tools in the age of social media. Ms Munchetty commented that, if invited in, the media can make good TV or radio, it just wants pictures, excitement and stories; nuclear energy is still a mysterious concept to the general public, popularized as being dangerous in movies and television.

Panellists also noted that, as a tight-knit industry, international cooperation has worked well to help stakeholders learn from previous accidents. International exercises such as ConvEx-3, as well as annual national exercises, are valuable for learning on and improving safety. The accident highlighted the challenge of the logistics involved in accepting offers of assistance and donations, and their distribution. Looking to the future, with SMRs and new technologies, the USA will likely use an all-hazards approach, with more resources for communication with the public to develop trust, and market its increased uses. In Brazil, EPR for SMRs will be based on source terms using a graded approach. All agreed that data from the industry for EPR that has been reviewed and validated by the IAEA is trusted. Employing the Global Positioning System (GPS) for movement and Twitter for social data was also suggested as useful. The nuclear industry needs to continue striving for being incident free, keep the conversation positive, and continuously improve safety so that the public trust and interest in nuclear energy as a safe, efficient non-carbon source continues to grow.

Questions from the floor demonstrated frustration with getting the message out in spite of a plethora of information available. The industry needs to use innovative approaches to get its story out. Social media training is one such avenue. Ms Munchetty noted that this is a brilliant industry but underexposed to the public. A lot of good learning has been done, now is the time to show the world how we can assist with the climate crisis. Communicating differently, concentrating on the safety and benefits of new advanced technology and its value in future applications would aid in removing barriers to acceptance.

5.3. PANEL 3 – PROTECTING PEOPLE AGAINST RADIATION EXPOSURE – ATTRIBUTING HEALTH EFFECTS TO IONIZING RADIATION EXPOSURE AND INFERRING RISKS

Ms M. Crane: Moderator
Mr T. Boal: Australia

Mr C. Clement: International Commission on Radiological Protection
Mr M. Kai: Japan/Nippon Bunri University
Mr W. Mueller: Germany/German Commission on Radiological Protection
Mr M. Tsubokura: Japan/Fukushima Medical University
Mr S. Zhaorong: China/Nuclear and Radiation Safety Centre

In the aftermath of the Fukushima Daiichi NPP accident, the radiation risk coefficients used for radiation protection purposes were not properly interpreted by the media and members of the public. While such coefficients derive from conjectures on potential nominal risks, they were used to make theoretical calculations of nominal effects that were attributed to the low radiation existing exposure situations resulting from the accident. At an international level, there were overestimations of the doses received by members of the public, which might have contributed to causing anxiety to the public.

The limitations for attributing radiation effects following low level radiation exposures need to be discussed and clearly explained. This panel discussed these perceived inconsistencies vis-à-vis the latest UNSCEAR estimates of attribution of effects and inference of risk, which became available after the Fukushima Daiichi NPP accident.

Ms Crane noted that it is difficult for the public to understand technical information such as risk coefficients and this can cause health effects such as anxiety or serious misunderstandings.

Mr T. Boal noted that we can use epidemiological models for high doses but are limited in our understanding of low doses.

Mr C. Clement noted that there is a great deal of uncertainty with respect to what dose causes, i.e. too low or long term exposure. (Was the cancer caused by radiation?) We have to be open to communicating what we know, what we don't know, and uncertainties; and within that context, decisions still have to be made. We have to apply both science (LNT and mSv) and ethics (benefit over harm).

Mr M. Kai joined virtually and reflected that it is not about a change to risk models but how to communicate clearly, quickly and concisely with the public.

Mr W. Mueller joined virtually and stated that it is easy to study high doses of radiation and explain the effects, it is much more complicated at low doses and stochastic effects.

Mr M. Tsubokura joined virtually and noted that the health impacts of the accident were much larger than the radiological impact. Mr S. Zhaorong joined virtually and suggested that now that 10 years have passed after the Fukushima Daiichi accident, we need to collect more data and combine it with that of Chernobyl to review and assess the effectiveness of the LNT model.

The panel discussed the effectiveness of the LNT model, noting that it can be used as a tool to explain matters to the public quickly and easily and serve as a model for practical solutions. However, LNT may be too simple to explain what is happening at various doses, which is much more complicated than a straight line, and overall health effects are not related to LNT. Panellists also considered what a reliable timeframe would be for studies on stochastic effects and what needs to be removed from the study, noting studies can go on for decades and we need to understand the mechanisms of the cancers or other long term effects and deal with screening effects, which can be serious. As for the issue of communicating uncertainty, panellists noted that there is no point in going into highly specific detail when all people want to know is if it is safe. We should answer the questions asked openly and honestly. In answer to a question such as "Can I pick mushrooms here?" we can perhaps provide information from the authorities or refer people to a service where they can have things measured.

Responses to questions from the floor highlighted that we need to respond and work with the public, catering our communication to the audience, using less jargon, and perhaps less precision than required by specialists. We can work with journalists who specialize in communicating with the public. In terms of medical communication, the aim is to provide assistance in a way that makes them comfortable. There may be a role for the IAEA in this area.

5.4. PANEL 4 – INTERNATIONAL COOPERATION

Ms H. Vaughan Jones: Moderator
Mr A. De Los Reyes: Spain/Nuclear Safety Council
Mr M. Garribba: European Union/European Commission
Ms C. Georges: France/French Alternative Energies and Atomic Energy Commission
Ms O. Lugovskaya: Belarus/Gosatomnadzor
Ms S. Udomsomporn: Thailand/Office of Atoms for Peace

The Fukushima Daiichi NPP accident emphasized the importance of international cooperation in safety related areas, including safe operation, emergency preparedness and response and regulatory effectiveness, and of incorporating lessons from the accident into international programmes to build capacity for more resilient systems.

Institutional networks for safety, such as regional networks, knowledge networks and regulatory fora, provide a platform for information exchange and help to optimize resources, compare processes, procedures and policies, identify good practices, and identify and address existing gaps and needs. The panellists shared different perspectives on how international cooperation contributes to establishing an international framework and global commitment for nuclear safety.

Mr A. De Los Reyes remarked that originally Spain's regulations were based on the technology that they had; it was a narrower perspective. Today they are more active internationally, which provides a broader view and more information on safety, as well as allowing them to assist those countries developing their technologies. He noted that international cooperation should be well panned and established to avoid duplication of effort.

Mr M. Garribba, joining virtually, noted that with a mix of older operators and more new operators arriving, it is important to share experience to learn from one another. The European Commission has made enhancements to their nuclear framework in response to the Fukushima Daiichi accident as well as the ensuing stress test.

Ms C. Georges, appearing virtually, provided an overview of France's assistance to Fukushima following the accident and how they learned from each other.

A copy of the points made in Ms Georges' introductory statement and those she wished to highlight in the panel discussion have been provided as follows:

I have been involved in different collaborative initiatives related to 1F since 2013. They took different shapes:

— Through workshops and final reports at IAEA ("DAROD: Decommissioning after a Nuclear Accident: Approaches, Techniques, Practices and Implementation Considerations") or at NEA ("EGFWMD: Management of Radioactive Waste after a Nuclear Power Plant Accident");
— Bilateral collaboration agreements with NDF, JAEA and TEPCO, with exchange of information and lessons learned in Marcoule;
— R&D collaboration for METI;

— Links with Japanese universities;
— Invitations to conferences, participation in European projects dedicated to decommissioning and waste management.

There are several points I want to emphasize:

— There are a lot of similarities between decommissioning of a reprocessing plant and Fukushima Daiichi reactors, specificities different from decommissioning of normal reactors due to presence of radionuclides: irradiation, contamination, criticality, waste management, etc.;
— Win-win collaborations with focus on opportunities for CEA and other French industrials in return;
— Rich human and cultural experience (not only technical; e.g. contaminated water, links with Iwaki community, etc.).

Points in the panel discussion:

— Zoom in on the unique way that the Japanese government early on considered the interest in innovative technologies from the international community through RFI / RFP (with contacts through national entities in charge of huge decommissioning programmes: DOE in USA, NDA in UK, and CEA in France – In 2013, I was the contact point for CEA) and how the information collected can be a rich source of knowledge for the decommissioning community. But difficulties for industry if no support from state company;
— Zoom in on how Japanese bodies were invited to European projects, as end users, etc.;
— Zoom in on EU project SHARE, which I coordinated, that will provide early in 2022 a road map for future collaborative work in the field of research (8 areas, not only technical, including safety, environmental remediation, knowledge management and training, etc.) for decommissioning with a proposal of instruments for implementation.

Ms O. Lugovskaya reflected that NPP accidents, such as Fukushima Daiichi and Chernobyl, show that safety is an international issue. She highlighted international cooperation in providing assistance for recovery efforts and provided an overview of Belarus' international projects.

Mr K. Mrabit provided a personal anecdote of his background in international cooperation.

Ms S. Udomsomporn provided an overview of Thailand's participation in regional and international projects.

Ms R. Weston joined virtually, noting that the Fukushima Daiichi NPP accident shook the nuclear industry to its core; however, it also showed the strength of international cooperation.

A copy of Ms Weston's opening statement has been provided as follows:

"Good morning and thank you for the invitation and my apologies I could not join you in person in Vienna for this important conference. I offer some opening comments through a lens of personal experiences and bringing to life some of the mechanisms of international cooperation, using Fukushima as a case study. I have been privileged to visit the site more than once and offer support as an international expert on the last two IAEA Reviews of Fukushima Daiichi NPP accident.

As the Chief Operating Officer for Sellafield site in the UK, I can see the future challenges looming for Fukushima with the sheer scale of the waste management, retrieval and storage infrastructure required to decommission the site, because of the parallels faced by ourselves in the UK of establishing waste routes and capabilities sufficient to deal with the volumes at the pace we want to work at.

Some events are seminal and truly industry defining.

On seeing and hearing of the accident, the unwritten international cooperation between operators all over the world sprang into action and we immediately organized to help gathering PPE and radiological instruments on a Saturday, loaded them onto a lorry and dispatched them to Heathrow airport on the Sunday.

Fukushima shook the nuclear industry to its core, with decisions across the world influenced by the events of March 2011. I spent my early career commissioning the mixed oxide fuel fabrication facility at Sellafield – The summer of 2011 the UK closed its MOX manufacturing facility and I remember briefing that news to that workforce.

After those initial weeks, as operators our focus turned to our own resilience and ensuring that we learned from Fukushima. The severity and the stark visibility of the impact of the accident to the world insisted the industry took action – and the existing international cooperation constructs supported in that – not least the work Mike Weightman and many others led and the role of the IAEA and the Action Plan on Nuclear Safety.

The rigour of the resilience work at Sellafield undertook in response to the EU Council of Ministers mandate of stress testing of nuclear facilities (that Massimo referred to) has moved us from a position of unwarranted confidence to that of healthy unease and a continued and renewed vigour to challenge ourselves and put nuclear safety as our overriding priority.

Also of note post Fukushima is that WANO also developed its Corporate Peer Review process worldwide as a result, and actually it caused Sellafield in the UK to seek the support of WANO in undertaking peer reviews for operations to retrieve legacy material, because of the potential consequence it poses to the environment, culminating in the first non-reactor WANO Pre-Start Up Review earlier this year.

As well as addressing our own gaps identified as a result of the lessons of Fukushima, the UK continued to support Japan via international collaborations, but also bilaterally. The UK and Japan have had longstanding relationships and nuclear expertise was seconded into the British Embassy.

It was here in Vienna at the IAEA that the bilateral non-commercial cooperation agreement was conceived between Sellafield and TEPCO in 2013 and signed in 2014 and renewed for a further five years in 2019, an obvious link being the experience of transitioning a workforce from operations to decommissioning and a site with a time imperative to deal with some of its legacy challenges.

As well as supporting technical activities such as spatial and long term planning and logistics, radiation protection and radiological roll back etc, environmental monitoring and safety management and Nuclear Safety Oversight Office, the human dimension was critical.

There was a real focus on communications and community support, leading to the Fukushima West Cumbria study and more recently the student exchanges as well as local government leaders and business cluster links. Sharing of ideas and secondment of people continues to grow to this day – and we are now mutually learning from each other in this regard. For instance, Sohda-san, a TEPCO secondee at Sellafield, said "I have spent 3 years seconded to Sellafield and have been able to gain some vital experience. Working as part of the Pile Fuel Cladding Silo Commissioning Team has enabled me to see first-hand the way a complex retrieval project is being brought into service."

This operator-operator agreement was followed in 2015 by a cooperation agreement between NDA and the newly formed NDF that helped the Japanese government establish the basis of a funding mechanism and a plan that recognized the lifetime activities required to remediate the site.

These agreements are in the spirit of cooperation with Sellafield learning from TEPCO as well as sharing expertise – for instance some of the Project Delivery made very rapid progress on facilities such as the structure to remove fuel from Unit 4 as compared to some of our delivery timescales – and there now exist joint programmes on robotics and remote technology that has also enabled and supported wider international and commercial partnerships to develop.

I have been an international expert on the last two Fukushima reviews and never cease to be impressed by the activities delivered, but recognize the challenges faced with regard waste management in its broadest sense – and is where international collaboration can provide valuable support in ensuring lessons are taken for risk based, rather than measurement criteria, waste management.

In summary, international cooperation matters:

a) It is an industry reality that there is a dependency of every nation's programme on the continued confidence in safety and regulation of the industry.

b) Drawing on international support and learning is a strength, particularly when addressing complex and unique problems.

c) A waste orientated focus should be a key emphasis for global nuclear challenges; including transition from operation to decommissioning and the fitness for purpose of solutions when time at risk as a key principle.

During the panel discussion, the panellists shared different perspectives on how international cooperation contributes to establishing an international framework and global commitment for nuclear safety, and what more could be done. They noted that international, regional and bilateral cooperation is vital for establishing mechanisms for provision of assistance and it is important to avoid duplication of efforts. Cooperation mechanisms established and agreed in case of emergency need to be in place for strengthening nuclear safety worldwide. All agreed that the international legal instruments and IAEA safety standards had evolved after both the Chernobyl and Fukushima Daiichi NPP accidents to reflect findings and streamline the direction for international cooperation. The IAEA plays a vital role in facilitating the independent and transparent exchange of experience, where all parties benefit. Cooperation in such areas as post-accident recovery, leadership, safety culture and capacity building has been a focus. They suggested that in nuclear safety, technical cooperation should be utilized for both developed and developing countries. International cooperation is needed and will be instrumental in avoiding the politicizing of nuclear safety.

Questions from the floor highlighted the strengthening of international frameworks, such as the European Commission regulations and the IAEA Incident and Emergency Centre following the accident, as it forced the community to re-evaluate how they thought about safety as well as international collaboration and systems. It was noted that, following accidents, there is an opportunity to create strong international networks for continuous improvement in leadership and capacity building. The IAEA can serve as a champion of international cooperation with respect to nuclear safety and security. We are confident in our international systems; the challenge is how to communicate in such a way that we spread that confidence."

5.5. PANEL 5 – SPECIAL YOUTH PANEL – YOUTH AND THE NUCLEAR INDUSTRY

Mr R. M. Grossi: IAEA Director General
Ms I. Illizastigui Arisso: Panel Moderator

Mr H. Rogers Page: Panel Moderator
Ms D. Dantas Sardinha: Brazil
Ms K. Graham-Shaw: United Kingdom
Ms D. Maheswarasarma: Sri Lanka
Ms I. Meniailo: Russian Federation
Ms N. Promprasert: Thailand
Mr T. Scott Smith: United States of America

The IAEA invited students and early career professionals up to 30 years of age to submit essays on selected topics related to the themes of the conference. The aim of the essay competition was to promote creative and innovative thinking and highlight the critical role that the next generation will play in sustaining and ensuring a safe future for the peaceful uses of nuclear technology in areas such as nuclear power, food and agriculture, water management, and human health. The essay competition attracted 250 submissions from 60 countries. Finalists were selected through a blind evaluation process and attended the conference to participate in this special youth panel. The themes from the winning essays, as well as the discussions during this panel, were considered during the development of the Call for Action.

The Director General introduced the panel members and asked each to summarize their winning essay. He noted the importance of having youth participate in the conference and recognizing their work. The youth of today will be responsible for global nuclear safety in the future and the essay competition was one of the best ways to engage and involve youth, supporting them and their education and giving them a platform to express their ideas and contributions. Their perspective is important. The panel discussed the future of power generation, changed by technology and influenced by non-technical factors, such as climate change, public trust, the need for carbon neutral power generation, and the need to build trust in the safety of the industry through communication, transparency and education. The media has a crucial role to play and there are many sources of information on the internet, but the information has to be clear, concise and verified, providing tangible evidence to the public. Non-nuclear power applications and new nuclear technologies were also discussed. The winning essays were as follows:

Mr N. PROMPRASERT, What Does the Future of Power Generation Look Like and What is the Role of Nuclear Power the Years Ahead?

In the essay, the author analyses the history of nuclear power and expresses the opinion on the future power generation with nuclear power as part of the national energy mix in Thailand, a newcomer country that has never operated NPPs before. He describes the synergy of nuclear power and other reviewable sourses of energy as the path for national energy development.

Ms D. DANTAS SARDINHA, Nuclear Energy for the New Generation

In the essay, the author analyses the role of media and messaging for perception of nuclear power as dangerous or safe by public. She highlights the importance of short, clear but correct messaging and data, as well as role of online media for modern population. She gives vivid examples of how presentation of information can change perception of that information, she highlighted the role of journalists and communications and that due attention to that is required from the side of nuclear community to allow for broader understanding and acceptance of nuclear power.

Ms K. GRAHAM-SHAW, The Future of Nuclear Power

In the essay, the author discusses trend in nuclear power generation through the controversial past, significalt decline and potential development for the future. She reviews its portion in global energy mix and analyses numbers and locations of reactors that approach the end of their lifetime as well as those under constructions. She reviews innovative reactor designs and presents this as an opportunity for the broader use of nuclear power.

Ms D. MAHESWARASARMA, (No title)

In the essay, the author highlights the importance of science communication and positive messaging on nuclear power for it to be accepted by people and communities as a viable option for a carbon free energy generation. The author notes that despite availability of shart messages, the knowledge needed to understand potential risks and benefits of nuclear power is often difficult to acquire. Means of outreach and communication are being discussed.

Mr TRAVIS SCOTT SMITH, Public Trust, Climate Change, a Changing Workforce, and Capital Investment: Non-Technical Matters and Their Impact on the Development of Nuclear Power Over the Next Twenty Years

In the essay, the author analyses regional and national factors that have an impact on increasing or declining role of nuclear power now and in the next decades. He highlights the lack of public appetite for nuclear power, a focus on renewable energy, and a poor investment environment and consequent decline of generating units in some Western countries and increasing energy demands and state investment in expanding nuclear capacities in Asia.

Mr ILIA MENIAILO, 20 Years After the Fukushima Accident. Attracting Young Specialists to the Nuclear Industry. Nuclear Energy Development

In the essay, the author pictures hypothetically the world in 50 years where the climate change is irreversible but nuclear power provides humanity with necessary energy in a safe and reliable way. He describes how currently young generation perceives nuclear power in the challenging years after the Fukushima Daiichi nuclear power plant accident and what is needed to improve that perception to allow nuclear power serving the humanity that will be facing natural and other challanges.

5.6. PANEL 6 – SAFETY FOR NUCLEAR DEVELOPMENT

Ms M. Crane: Panel Moderator
Mr A. Bolgarov: Russian Federation/ROSATOM State Atomic Energy Corporation
Mr F. Dermarkar: Canada/CANDU Owners Group
Mr S. Ghose: Bangladesh/Bangladesh Atomic Energy Regulatory Authority
Mr L. Ma: China/National Nuclear Safety Administration
Mr L. Mlynarkiewicz: Poland/National Atomic Energy Agency
Mr S. S. Ata-Allah Soliman: Egypt/Egyptian and Radiological Regulatory Authority

This panel discussed the responsibility shared by the IAEA and Member States to ensure sharing nuclear technology is done in a way that adheres to the highest standards of nuclear safety, security, and non-proliferation. Currently, several countries are embarking on new nuclear power programmes, while others are expanding their existing uses of nuclear or radioactive material for industrial, medical and research purposes. The panel examined how vendors, recipients and international organizations can all play a role in ensuring that nuclear safety remains a global priority, and highlighted the role of a robust nuclear safety infrastructure and system in enabling future nuclear development, building on INSAG-27.

Mr A. Bolgarov provided the perspective of a vendor country, providing technologies to Russian codes for foreign customers. These codes have been compared to SSR-2/1 and SSR-2/2 and proved to be in line, as the Russian safety framework is aligned with the IAEA safety standards. They also compared a small land based SMR with IAEA safety standards and will build one in Russian

Federation to demonstrate safety to customers. Dialogues between operator and regulator are well established.

Mr F. Dermarkar appeared virtually, and also from a vendor country, noted that there is no path to net zero without nuclear energy. Multilateral cooperation is key, and it is vital for established countries to share experience, as many states will become nuclear operators in the future.

A copy of Mr Dermarkar's remarks has been provided as follows:

It is an honour for me to be part of this panel today, and I would like to thank the IAEA for hosting this important discussion. Net zero by 2050. This sentence is possibly one of the most repeated in the world today. And most of us gathered here today recognize that there is no path to net zero without nuclear. Many countries that are not operating nuclear reactors today will become operators in the near future. And that's what makes this discussion today so relevant.

Before we get into the discussion, I would like to put a few thoughts on the table.

First, it is vital for countries with established nuclear programmes to play a leadership role in sharing their knowledge, their experience and their competence to smooth the path to net zero for embarking countries.

Second, we currently have 450 or so reactors in about 30 countries. It is important to recognize that the international and national frameworks for ensuring their safe operations will need to adjust as the number of operating reactors reaches into the thousands across more than 100 countries. We may need to put elements into place internationally that do not exist today. This notion of evolution of international frameworks is not new. For example, WANO, which has become one of the key international organizations for assuring nuclear safety, was not established until 1989, almost three decades after the birth of commercial nuclear power production.

Third, when establishing national and international frameworks for nuclear safety, it is important to consider the entire lifecycle of the nuclear installation, from site selection through to final decommissioning and waste disposal. There is much focus today on harmonization of design so that a reactor that is licensed in one country can be readily accepted by another. There is no doubt that this is essential. But equally essential is the need to maintain a harmonized design and licensing basis as the reactor ages and lessons are learned from operating experience.

We are gathered today to recognize the 10th anniversary of the Fukushima Daiichi NPP accident. One of the learnings from this accident is that important design changes to mitigate severe accidents that were made by operators in some countries did not find their way to reactors of the same design at Fukushima. If nuclear technology is to get adopted by more countries, it is essential that we develop a framework that would ensure that relevant operating experience and research findings for a given reactor design are implemented consistently for reactors of the same design, regardless of the jurisdiction in which they operate.

All of this ultimately speaks to the importance of multilateral cooperation and collaboration in sustaining nuclear safety. I would like to recognize the essential role that several key organizations play, including the IAEA, NEA, WANO, WNA and the various reactor owner groups. These organizations have a key role to play in helping to chart the path forward that will enable widespread deployment of nuclear technology while ensuring nuclear safety. The frameworks they have established for sustaining nuclear safety are necessary, and they need to be re- examined to ensure they continue to be sufficient for assuring nuclear safety as many more countries and operators adopt nuclear energy as a solution to climate change.

Mr L. Ma was unable to connect virtually, however a copy of his remarks has been provided as follows:

China treats nuclear safety as an important obligation of the state, and exercises unified regulation through special organizations and a regulatory system underpinned by independence, openness, the rule of law, rationality, and effectiveness. To ensure independent regulation of nuclear safety and enhance its authority and effectiveness, China has strengthened technical support and developed a professional team while modernizing the system and the regulatory capacity.

In China, unified regulation over the surveillance of nuclear safety, radiation safety, and the radiation environment is exercised independently, and a three-pronged regulatory system consisting of headquarters, regional offices and TSOs is in place. Local governments at all levels regulate regional radiation safety through regulatory organizations or full-time/part-time regulators according to local conditions.

The peaceful development and utilization of nuclear energy is the common aspiration of all countries, and ensuring nuclear safety is their shared responsibility. China advocates the development of an international nuclear safety system characterized by fairness, cooperation, and mutual benefit. It facilitates the global effort on nuclear safety governance through fair and pragmatic cooperation, works together with the rest of the world to build a community of shared future for global nuclear safety, and promotes the building of a community of shared future for humanity.

China attaches great importance to nuclear safety policy exchanges and cooperation between countries. It maintains close contacts with France, the United States, Russian Federation, Japan, the Republic of Korea and other countries, as well as emerging nuclear energy countries in the framework of the so-called 'Belt and Road Initiative', and has signed more than 50 cooperation agreements on nuclear safety to facilitate all-round cooperation in exchange of high level visits, communication between experts, review, consultation, and joint research.

China has strengthened exchanges and cooperation with the Nuclear Energy Agency of the Organisation for Economic Co-operation and Development, the European Union, WANO, and other international organizations. It is an active participant in international peer reviews of nuclear safety directed to common progress against global standards. In order to expand participation in global cooperation platforms and enhance its nuclear safety capabilities, China continues to take part in activities under the frameworks of the Global Nuclear Safety and Security Network and the Asian Nuclear Safety Network. China contributes its wisdom and strength to the world by promoting its nuclear safety regulatory system and sharing advanced technology, experience, resources, and platforms. It has taken part in the Multinational Design Evaluation Programme for NPPs, and a working group on the Hualong One reactor (also known as HPR1000) has been established. Through its National Research and Development Center for Nuclear and Radiation Safety Regulation, China has continued to help developing countries to train nuclear safety personnel and carry out technical drills, lending support to their efforts to enhance their regulatory capacity and providing more public goods for improving global nuclear safety.

China's first NPP began operation in 1991, and by 2030, the country, plans to have about 90 NPP units either in operation or under construction. To support this rapid expansion, China has been developing its own nuclear power technology. China is also starting to export this technology to countries embarking on nuclear power programmes. The government of China is strongly committed to assisting countries strengthen their regulatory framework in the implementation of this technology. To be able to better assist regulatory bodies of embarking countries that intend to use this technology, MEE (NNSA) requested a policy discussion on the best methods to support these regulatory bodies.

Mr S. Ghose provided the perspective of a new operator, noting that safety is not a product but a prerequisite. The challenge is how to develop nuclear infrastructure so that safety is a prerequisite. IAEA peer reviews and advisory services help to strengthen infrastructure and safe and sustainable technology has been chosen with the support of the vendor. The vendor country can also help not only with the financial challenges but with the human resources challenge.

Mr L. Mlynarkiewicz contributed as a recipient country, noting that safety is not a one-time issue, NPP licensees and regulators need to seek out new information that affects safety. Regulators have to be ready to also oversee new nuclear development. Capacity building is key as is a technology neutral system. Human, financial and infrastructure resources as well as organizational development are challenges. It is important to be a knowledgeable customer, especially about new technology.

Mr S. S. Ata-Allah Soliman provided the background to Egypt's nuclear development and its current objectives as a recipient country with the support of a Russian TSO. He stressed that legal and national frameworks need an integrated management system and an independent regulatory authority with governmental financial and resource support.

Panellists also discussed the international potential for mutual recognition of safety assessments, as well as the concept of an international TSO created under the auspices of the IAEA.

The panellists discussed the expected expansion of nuclear power use, noting that safety is a prerequisite to nuclear power generation; it has to be built into the supply chain. Multilateral cooperation and collaboration are vital. Excellent project management is important. For a new nuclear power project in a country that does not have its own design, the main safety challenges to consider before starting to look seriously at nuclear power include a need for appropriate independent regulatory structures and the development of institutional and human resource capacities, using IAEA safety standards and advisory missions, to strengthen national systems for safety; infrastructure development in an enabling environment with cooperation between the vendor and key stakeholders; the building of national legal and regulatory frameworks with integrated management systems; knowledge management and transfer, as TSOs and vendors assist in capacity building. Safety is the responsibility of the operating organization, but the regulatory body plays a vital role in ensuring that there is adequate financial support, and that competence is developed and maintained.

They discussed the challenges in ensuring the national competence for safety and how to work together with technology providers and the concept of the responsible technology vendor: what this means in reality, and any changes that occurred after 2011. Attributes of a responsible vendor include: the vendor provides support for the entire lifecycle of the facility, does research to stay one step ahead, collecting and sharing design and operating experience, and ensuring its implementation; the vendor builds national responsibility and provides for the orderly transfer of intellectual property if and where needed; there is a mechanism to transfer knowledge to successor operators. The owners group works together with the vendor. The current framework needs to be re-examined for sustainability. With the advent of SMRs and advanced reactor technologies, international and national frameworks for safety will have to evolve.

The panellists also discussed whether there is international potential for mutual recognition of safety assessments for different technologies from different vendors. Accountability is on the national level and cannot be delegated, but others' safety assessments can provide information. They agreed that there is room for sharing information. This may be different for SMRs, as designs are meant to be manufactured and shipped to a country. The SMR Regulators' Forum is working to develop a process for regulators' cooperation, recognition of other regulators' reviews, legal constraints, etc. with the recognition that there might be fabrication in series for SMRs. It was suggested that perhaps a regional generic design assessment would be useful, or the creation of an international TSO under the auspices of the IAEA. It was noted that interfaces between the stakeholders are complex and diverse, and that it is important to have government to government, and regulator to regulator, agreements, as well as agreements between TSOs, industries, universities, etc. All agreed that there can be no compromise on safety but budgeting for all safety related upgrades and changing requirements as a newcomer country is difficult. There is a need for adequate funding for the regulatory body.

In the Q&A session that followed, the need for more information on research and development, maintaining the pipeline for future operators, more support for existing international programmes, and more emphasis on the important role of the TSOs were highlighted.

5.7. PANEL 7 – BUILDING INCLUSIVE SAFETY LEADERSHIP

Ms H. Vaughan Jones: Panel Moderator
Ms G. Gudela: Switzerland/Eidgenössische Technische Hochschule Zürich
Ms M. Lacal: United States of America/Palo Verde Generating Station for Arizona Public Service Company
Mr N. Maqbul: Pakistan/Pakistan Nuclear Regulatory Authority (PNRA)
Ms E. Romera: Spain/Nuclear Safety Council
Mr C. Viktorsson: United Arab Emirates/Federal Authority for Nuclear Regulation (FANR)
Mr B. Zronek: Czech Republic/CEZ Group (CNO/CEZ)

Global and national labour markets are changing. Personnel with difference mindsets, different expectations and different competencies are joining the nuclear sector. Nuclear operations represent a complex and dynamic sociotechnical system that benefits from a systemic approach to safety and proactive leadership and management to assure safe performance. Reliability of human resources and availability of competence for safety depends on multiple factors. "Effective leadership and management for safety must be established and sustained in organizations concerned with facilities and activities that give rise to, radiation risks" is stated as Principle 3 of the IAEA Fundamental Safety Principles. The purpose of this panel was to draw attention to the changing workforce demographics in operating organizations, regulatory bodies, and TSOs. Panellists discussed strategies to address potential challenges and examined how nuclear sector practices are keeping pace with other industries to have an inclusive workforce and safety leadership in the mid-21st century.

Ms Gudela noted that there is always a need for a social science input to leadership and culture. Trust cannot be forced but should be gained. Research and development are necessary to ensure safety in innovation.

A copy of the summary of Ms Gudela's introductory remarks has been provided as follows:

Nuclear power and its future need to be considered as part of the global development of energy production in times of a major climate crisis. The clearer leaders can position nuclear power in this global picture the more convincing they will be in the eyes of younger generations of workers and the general public. To achieve this, transparent and holistic risk assessment and management is necessary based on an inclusive culture of interdisciplinary appreciation which heeds technological as well as social knowledge. Safety does not preclude but in fact has to rely on innovation; to ensure learning and change in response to external and internal challenges, leaders need to encourage worker engagement which is also key to keeping workers motivated and involved.

Ms M. Lacal joined virtually, noting that it is clear that NPPs play a role in providing clean energy, and the focus needs to be on maintaining the safety of the existing fleet, and that one needs to openly challenge and receive feedback on performance.

Mr N. Maqbul noted that leadership and safety culture are very important. Soft issues such as culture, are equally as important as technical aspects, such as defence in depth. PNRA has a well-established leadership programme.

Ms E. Romera noted that accidents have changed the perception of safety and that sharing experience is essential.

Mr C. Viktorsson provided the background to the United Arab Emirates' nuclear programme, which included bringing in foreign experts to lead development. The United Arab Emirates performed a stress test after the 2011 accident.

A copy of Mr Viktorsson's introductory remarks has been provided as follows:

As regards Fukushima, I would say that in the UAE it made an alarm ring, it showed that nuclear is different! It is not oil and gas! The mindset turned to the seriousness needed in dealing with nuclear. When you start the fission process, you are obliged to manage the active core for long time whether it is producing or not producing electricity! So the seriousness of nuclear business became evident, and I think it is a strong reminder for the nuclear sectors in all countries that we have to continue to fight complacency. Also, to continuously question ourselves if we have identified and corrected deficiencies or if not do we have plans to do so.

1. I think we can agree that the role of nuclear leadership is not discussed enough when we meet internationally on nuclear safety, more is needed. That's why I think our panel is of crucial importance. I think we need to talk about leadership in general in nuclear industry and regulation, not only the role of the top leader but the entire chain of leaders in regulatory bodies and industry. I have clearly seen a need to address middle management in a way allowing them to be able to communicate the vision, values and expectations from top management to staff in a way that is clear and understood. Also, the behaviour of staff needs to be continuously monitored and the desired values need to be enforced. We have joined forces with Nawah in a leadership training programme that will address all types of nuclear leaders. It is of special importance in our situation as newcomer and with so many different languages and cultures working in the nuclear sector in the UAE.

Visible and clear vision, values and expectations as well as follow up to enforce that expectations are met. Safety has to be the overriding priority, not the schedules.

2. For the same reason, we also see both at FANR and in the industry the importance to develop the "FANR way" for our regulatory approach and the "Nawah way" within the industry. My message is clear to staff, "Make sure you are convinced before you approve". The Nawah's lead concept is "we move when we are ready", not blindly follow the project schedules. These concepts have worked well and have not really delayed the project but rather given us many opportunities to stop and reassess before moving on. It relates very well to the team of the conference of "there is no place for complacency"! Move when we are ready!

3. As I mentioned above, we have so many different cultures, languages and traditions within our organizations, at FANR around 30 different nationalities, and even more at Nawah. It was evident for me from day 1 that this matter was going to be a challenge, and that we needed strong management systems for safety to support the leadership. Consequently, the first regulation we issued was on leadership and management for safety. We largely followed GSR Part 2 of the IAEA. At FANR, we introduced our integrated management system in order to ensure we all act "the FANR way". It has been clear from all of us, both at FANR and at ENEC/Nawah, that the integrated management system is a key to support the leaders work to promote a strong safety culture and ensure continuous development of needed skills. Foster a homogeneous language and strong organizational culture based on shared values. Establish a solid foundation for a strong culture of safety.

4. I think what we also have realized more clearly after Fukushima is the importance of safeguarding the independence of the regulator, but also that industry leaders have a clear vision where they are heading also in terms of safety based on continuous reassessment of safety. Our way is to have frequent top management meetings between myself, my deputies and the CEO/CNO of the applicant/licensee to ensure we are addressing the "right" issues at the right time, but also that we develop a road map for enhancement that we see as needed but might not be urgent, and for handling the unexpected. A strong operating experience feedback system is a must to keep high attention to the uniqueness of nuclear. The business continuity planning showed us how we could manage a crisis during the ongoing pandemic, for example. So, flexibility and agility are essential. Mutual respect of roles and responsibilities are prevailing in the constructive dialogues we have, and which is part of the success of the UAE nuclear programme. Understand what responsibility for safety means, i.e. not only to comply with regulatory requirements. A nuclear leader needs to look beyond the present and consider the future and the need to look for enhancements. Leaders have to guard against complacency and support business continuity management plans as well as care for high quality.

5. Of course, an important role for the nuclear leaders is to ensure we have the right competence to be able to operate/regulate the power plant safely and reliably. Here we come the question of inclusive leadership and decision making. We need the technical competence now and in the long term. For us the challenge is to build local expertise and leaders, and we have so far succeeded very well, at FANR almost 70% of the staff are locals, 40% females. Nevertheless, in some areas like nuclear safety we will need foreign experienced workforce for many years. Another aspect is the diversity of opinions to enrich the decision making and make it more solid. The multidisciplinary approach that is coming out of the Human Technology Organization (HTO) concept is so critical. Also, the need to allow outsiders to assess our work. We have at FANR systematically used IAEA peer reviews, and WANO peer reviews at Nawah. Also, various advisory groups have been seen to contribute value to our work. Leaders who see the value of openness to diverse views and welcome external views and assessments. Base decision on a systemic perspective

6. One particular strength we have in the UAE is the strong support we have from the government. The most essential parts of the support was the Nuclear Policy from 2008 and its principles and the commitment the government made to apply best international practices and join all essential nuclear conventions. This policy paper, which is of relevance to FANR and ENEC/Nawah in particular, but also to other stakeholders in the country, is still guiding us in our work. Those principles were given before we started to build FANR, and ENEC but were of great value not only in building the institutions, but also in attracting foreign experts due to the fact that they built on best international practices, and thus appealing to safety conscious experts. It is important to emphasize that the government or any stakeholders never have tried to influence any decision or timeline we have established, but they are kept updated on progress and are always supporting what FANR decides.

The nuclear governance gets solid if it has the strong support from government in terms of safety, security and nuclear non-proliferation principles and expectations from government at highest levels.

We need to guard against system deficiencies in the governance which could be indicative of poor safety management and poor safety culture in both industry and government!

Additional background information from FANR is contained in Annex V – Supplementary Files.

Mr B. Zronek emphasized that leadership for safety is one of the most important aspects of nuclear safety.

A copy of Mr Zronek's opening remarks has been provided as follows:

Leadership for safety can be recognized as one of most important issues not only of the Fukushima Daiichi accident. We can find similar attributes also in the Three Mile Island and Chernobyl accidents.

We can also find many other examples both within and out of nuclear industry, e.g. Deep-Water horizon event.

In this regard, it is highly desirable to keep reminding these events regularly and share with coming generations – not only to remember and get lessons learned but especially to keep the emotion and experience. This will prevent our dilution of awareness and attentiveness to safety related issues.

Our leadership for safety is the key part on the way to success even in current economically driven world. Nuclear operators are very often facing high expenditures and economic pressure, but we have to keep in our minds what the golden rule is: Only the safe and safely operated plant can be profitable in a long term perspective.

The panel discussed the main challenges that could have a negative effect on the nuclear industry and its safety and leadership approaches, in particular, guarding against complacency and being open to other perspectives. Other challenges were noted, such as national capacity building, using an integrated management system to integrate multi-cultural workforces under one cohesive organization; the integration of safety and security as a single culture, harmonizing safety culture concepts and working towards making nuclear safety culture universal, to ensure that important messages across cultural barriers (the IAEA can help by providing tools and advisory services); maintaining a consistent approach to safety culture throughout the supply chain and across all stakeholders to avoid the potential pitfalls of interpreting 'safety culture' (there should be only one definition and the key is sharing values to increase margins of safety with authenticity and transparency). The challenge of finding the correct balance of adaptive vs procedural approaches for and within the nuclear industry and the meaning of a "questioning attitude" were discussed. The ability to learn from other industries was highlighted, such as incorporating a mix of reflexive and reflected learning as in the aviation industry. The nuclear industry's challenge is the incorporation of more reflected learning, as this contrasts with adherence to procedures.

The role of the regulatory body or government institutions in ensuring that safety is the overriding priority was discussed. Regulatory bodies need to provide an example for operating organizations. A listening attitude is as important as a questioning attitude, and leadership needs to provide a clear vision, values and expectations and follow up to enforce those expectations, in order to establish a strong foundation for a culture for safety. Typical issues in leadership and safety culture areas for plants in long term operation were also noted, as the number of plants continuing past 40 years is increasing. A strong focus on reducing uncertainty was suggested, i.e. clarifying the decision making process to shut down a plant and training staff for other opportunities, e.g. decommissioning. Knowledge management and knowledge transfer are crucial to an inclusive workforce, especially if the lifetime extends beyond 40 years. Continuity planning is key as the challenge of obsolescence is real. The panel agreed there has been a 'brain drain' from the nuclear industry since, in more mature countries, many people have retired. A leadership focus on succession planning and retaining talent is important. The COVID-19 pandemic highlighted the importance of safety culture, rules need to be followed, with no compromises in safety. It was noted that the 9th Review Meeting of the CNS scheduled in 2020 has been postponed, which has impacted the sharing of experience among the Contracting Parties.

Responses to questions from the audience highlighted that being open to outside views is important for good leadership and influences good safety culture in all industry stakeholders and that perhaps the IAEA should have more guidance on the influence of national infrastructure on fostering safety culture. The Chair summarized the session by noting that there is a need to harmonize guidance on safety culture.

5.8. PANEL 8 – INTERNATIONAL LEGAL INSTRUMENTS

Ms M. Crane: Panel Moderator
Mr D. Dorman: United States of America/Nuclear Regulatory Commission
Ms D. Drabova: Czech Republic/State Office for Nuclear Safety (appeared virtually)
Mr N. Ichii: Japan/Nuclear Regulation Authority
Mr R. Jammal: Canada/Canadian Nuclear Safety Commission
Ms A. Müller-Germanà: Switzerland/Swiss Federal Nuclear Safety Inspectorate
Mr B. Tyobeka: South Africa/National Nuclear Regulator

Over the past four decades, several important international conventions and other international legal instruments have been adopted and progressively strengthened to achieve and maintain a high level of nuclear safety worldwide. International conventions are complemented by national policies for safety and regional agreements. Other initiatives also form part of the ongoing international effort to strengthen nuclear safety, such as the Vienna Declaration on Nuclear Safety that was unanimously adopted by the Contracting Parties to the CNS. The IAEA plays a critical role in maintaining the international framework for nuclear safety. The establishment of the IAEA safety standards through an international consensus process assists in the harmonization of nuclear regulations and helps States to comply with existing international legal instruments. This panel discussed how the effectiveness of the international legal instruments for safety can be further enhanced.

Mr D. Dorman provided a context for the legal instruments, noting that international legal instruments are tools to assist us. Safety is the responsibility of the operator, and that the Conventions allow one to check on the progress of lessons learned.

Ms D. Drabova participated virtually. She noted that nuclear safety has improved significantly, but it should always be seen as a work in progress. The industry is going through change, which provides an opportunity to reinforce roles and responsibilities, and adherence to conventions.

Mr N. Ichii reiterated that the Contracting Parties to the Conventions perform a peer review process, to extract important good practices that are sometimes difficult to identify. The spirit of conventions is to ensure a high level of safety worldwide.

Mr R. Jammal stated that the aim was to get nuclear safety to carry the same weight as Safeguards at the Board of Governors. He highlighted that operators and regulators are not signees for legal instruments, but governments are.

Ms A. Müller-Germanà noted that much work has been done to improve the effectiveness of conventions, including the country review report. The Vienna Declaration on Nuclear Safety – a result of a Swiss amendment – is now an integral part of the CNS review process.

Mr B. Tyobeka noted that South Africa has signed onto many conventions and codes, and the country reports are always made available. Conventions assist in furthering nuclear safety, acting as a robust peer review.

In their discussion panellists noted that contracting parties of conventions perform a peer review process. The conventions are not applied directly but are reflected within national laws. The 7th review meeting of the CNS, held in 2014, was the first after the IAEA report on the 2011 accident. Outcomes included an emphasis on the importance of government support to the CNS process and an improvement in transparency, as all reports were published. They noted that the 8th and 9th review cycles for the CNS have been merged and that the role of Safety Conventions and Codes of Conduct in the continuous improvement of global nuclear safety seems to rely on modifying rather than drafting new instruments. The Vienna Declaration on Nuclear Safety, consensually adopted, enhanced safety and has become an integral part of the CNS and should be recognized as such.

Suggestions for enhancement included using more modern communication tools to allow greater participation without the need to be present in Vienna (or another centre), which might increase the effectiveness of existing legal instruments Countries are very willing to engage with the CNS process, but some contracting parties relegate this process below other pressing matters. It is sometimes difficult to be in Vienna for two weeks. Another suggestion was to review what the benefit to countries should be vs how many good practices and recommendations are obtained, as it is very intimidating for smaller countries to produce national reports. Using different types of not legally binding instruments to aid effectiveness, such as the Code of Conduct, was also suggested, as Conventions are difficult to change. Differences in implementation between nuclear and non-nuclear countries were also discussed. All agreed that the effectiveness of the IAEA safety standards increased after the 2011 accident; however, there will always be room for improvement.

Questions from the floor raised the following topics for consideration: The importance of conventions, and whether they are a regulatory instrument; the difficulty in changing them; the possibility that other, different instruments could be used to aid effectiveness, such as codes of conduct; the goal of standardization compared with the sovereign need for national responsibilities; how to review the legal instruments without giving the impression that safety is not strong enough and how might the multiple international legal frameworks work better together.

5.9. PANEL 9 – COMMUNICATION, ENGAGEMENT AND TRUST BUILDING

Ms H. Vaughan Jones: Panel Moderator
Ms S. Bilbao y Leon: World Nuclear Association
Ms P. Harvey: United Kingdom/University of Manchester
Mr J.-L. Lachaume: France/Nuclear Safety Authority
Ms P. Lucio: Spain/Nuclear Safety Council
Ms L. Sauer: Canada/Canadian Safety Commission
Mr V. Titov: Russian Federation/ROSATOM State Atomic Energy Corporation

Safe operation of nuclear facilities is ensured through the cooperation of multiple stakeholders, including operating organizations, regulatory authorities and TSOs. Interfacing with decision makers and professional organizations, as well as communication with the public through local communities and the media, needs to be transparent and clear. Availability of information from different sources and access to new formats of communication present opportunities but also challenges for conveying accurate and reliable information to decision makers, the media and the general public during both normal operation and emergency situations. This panel discussed how information can be shared accurately and in a timely manner in a way that is understandable to the target audience and builds trust with the public.

Ms S. Bilbao y Leon noted that most often, communication is an afterthought in the nuclear sector. Communication experts, social scientists, etc. have key roles in nuclear. Nuclear community does a fantastic job in communicating risks, but in needs to be admitted and addressed that it is not so good at communicating the benefits.

Ms P. Harvey joined virtually and noted that safe operation is enabled through many stakeholders.

A copy of Ms P. Harvey remarks has been provided as follows:

There is a need to foster an awareness of how complicated the interactions between stakeholders are. The public themselves are incredibly diverse, and not a singularity. We need to develop familiarity in communication.

This panel starts from the premise that safe operation of nuclear facilities is ensured through the cooperation of multiple stakeholders, including operating organizations, regulatory bodies, TSOs, decision makers (who will include political authorities at various levels: international, national, regional and local), expert bodies and the general public.

I would stress the need to foster an active awareness of the complexities involved in communication, engagement and trust building across this diverse field of interests, values, experiences and expectations. While there are well-established principles for effective engagement, these principles do not in and of themselves ensure effective communication or increased trust between interlocutors. For example, the International Association for Public Participation (IAP) has outlined the key stages of building effective public engagement that move from a commitment to deliver balanced and accessible information, through processes of consultation and involvement, to the creation of active collaborative partnerships and ultimately the empowerment of the public as decision makers.

To mobilize these principles as specific actions, it is very important to remain aware of the fact that there are many diverse 'publics' who take form around competing fields of value. Publics are political collectives – neither homogeneous nor preformed. They come into being in response to the consequences of the actions of others, actions that elicit responses of support, of influence, and/or of protest. Those charged with communication and engagement in the nuclear industry need to be aware of how others are reading their actions – and indeed misreading them. They need to remain curious about how specific initiatives are received. Given the long timeframes involved this will never be a once and for all moment of effective communication – even when such moments of engagement are achieved.

If we are to consider 'how information might be shared accurately and in a timely manner', we have to take into account that communication is a two-way process. To think about whether information is understandable to a target audience requires an understanding of who exactly the target audience is. If the aim is to build trust with a specific audience, then it is important to invest in engagement and communication as a reciprocal, two-way process. Who is the target audience? Is there only one? How well do we know that audience? We should not assume the coherence or homogeneity of social groups. We need to learn more about how people take decisions in everyday life with respect to other hazards and risks in their immediate environments. We need to know what matters to people – especially when what matters does not fit with own framings and understandings of the situations that we are trying to influence or communicate to others.

Familiarity is the basis of trust and good communication. Familiarity involves an ongoing learning process. Many things were learned over the past ten years. Unfortunately, these include an awareness of one-sided conversations – and a failure to find ways to appreciate and act on the qualitative aspects of nuclear safety. These might include a need for corporate actors to appreciate that a risk informed orientation to safety (as statistical probability) does not amount to an experience of protection or care; or for governments to understand that clear and decisive leadership on behalf of local populations also involves taking responsibility for people's well-being beyond the statistics of life and death; or for all involved to acknowledge past failures and to demonstrate the learning that such acknowledgement entails. Above all, it is important to remember that neither technical design nor communication strategy can offer guarantees of future safety. Safety can be enhanced by mutual understandings of known hazards.

Mr J.-L. Lachaume noted that there are no regulations regarding communication and trust building. Trust needs to be established in advance, not simply after an accident.

Ms P. Lucio noted that the role of communication is difficult for a regulator to gauge. NEA's Working Group on Public Communication of Nuclear Regulatory Organizations (WGPC) has been implementing useful activities.

A copy of a brief contents of Ms Lucio's remarks has been provided as follows:

- Introduction about topics related with the role of the regulator concerning communication and information management and how it impacts on the risk perception by public;
- A vision of the international regulatory environment related to the information and communication function;
- Outcomes from the accident analysis focused on the information and communication issues;
- The CSN legal framework regarding information and communication;
- The implementation of the Spanish National Action Plan after Fukushima concerning information and communication and other associated improvements;
- Thoughts on information and communication as a safety function skill, for debate and conclude.

Ms L. Sauer noted that the way people think about risk is complex. There are clear parallels between nuclear risk perception and vaccine hesitancy. We need to think of ways to bring stakeholders into the decision making process.

A copy of Ms Sauer's opening remarks has been provided as follows:

Hello everyone and a big thank you to the organizers of this panel for having me today. One of the many lessons learned from the Fukushima Daiichi NPP accident and re-enforced again and again during the COVID-19 pandemic, is that the confidence and trust in government, including regulators, is absolutely crucial to the public good. Why should trust matter to nuclear regulators? According to the OECD, trust is the foundation upon which the legitimacy of public institutions is built and is crucial for maintaining social cohesion. Without trust, regulators cannot maintain the confidence in their activities and decisions. While this is important on a day-to-day basis, this is particularly true during an emergency. Being a trusted regulator will increase the likelihood that the public will accept and comply with protective actions and other response actions in an emergency. The public needs to trust the information they are reading or hearing, so that they act on the nuclear regulator's advice and direction to keep them safe.

But building trust is an uphill battle: Trust in institutions is generally declining in most countries. According to the OECD, only 45% of citizens trusted their government in 2019. The public is no longer dependent on policy-makers or regulators for information – they turn to whom they trust the most – family, friends, influencers, select social media, etc. And to compound this, we are seeing misinformation circulate at an unprecedented rate. Current tactics to build trust, the ones we are using today, are not sufficient. Let's start with some of the beliefs that have driven many of our efforts:

1. That we are the experts, so people should believe us.
2. That if we share the data, what it means, people will believe us.
3. That if we could just find the most compelling facts, then people would believe us.

What have these beliefs resulted in? We have become very good at:

— Outreach = telling people things
— Transparency = sharing information
— Emphasizing independence and the scientific basis of our decisions

This is good, but not enough. Why? Decades of research have demonstrated that the way people perceive risk is very complex and is influenced by a wide range of factors – affective, cognitive, contextual, and individual. The way people think about risks can therefore act as a shield, preventing them from being open to and receiving messages about risks.

Let's consider this pandemic and vaccines. Billions of dollars were invested to develop vaccine candidates and prepare for global vaccination programmes to immunize against COVID-19 infection. Several vaccines have received regulatory approvals and been rolled out. The timelines – measured in months, not years – were astonishing. This has been an enormous achievement for science. This should have been a huge boost for public trust in science and government, and yet we are seeing important levels of vaccine hesitancy. Despite the science, a sizeable number of people are reluctant to get the shots. There are clear parallels between vaccine hesitancy and nuclear phobia. Both are proving stubborn to defeat, and both are immune to purely scientific arguments. "More and better" communication has limits. In other words, continuing to do all the great things we are doing now is simply not enough. We need to think differently.

But there is good news – trust is the ingredient that will help us. One powerful way to build trust is through true stakeholder engagement. And this is built on genuine two-way dialogue and the empowerment of stakeholders. How can nuclear regulators build true stakeholder engagement? We need to invest internally and externally. Internally: We need to build an employee culture that is truly supportive of increased engagement and trust building. This cannot be viewed as a distraction from our "real jobs". It needs to be understood as part of our core business. And it takes time and costs money. This means:

— Ensuring staff have the ability and behavioural competencies to engage and build relationships with all of our key stakeholders;
— Properly planning and resourcing engagement and trust building initiatives;
— Humility and agility – Governments and regulators had to change course frequently during the pandemic, and occasionally to share information that seemed to contradict what they had said before. The right approach there – one that we saw succeed in Canada more than once – is to own that. To admit to the change of position, and to embrace the humility of science – that our understanding is always provisional and needs to change when new facts emerge. This will be particularly true during any emergencies.

Externally: Start reaching out to your key stakeholders. Open a dialogue and accept that different stakeholders probably need different messages. Listen carefully and be prepared to adapt and tailor your approach. Look for ways to bring different stakeholders into your decision making processes. For example, in Canada, we have established a forum with our key environmental NGOs. We are attempting to build a dialogue so that we can understand each other better.

Once we have succeeded at meeting our stakeholders on their terms, in their forums of choice, they will be empowered. To maintain and further these relationships, we need to communicate openly, and in plain language with them, and frequently. In closing, we have one great advantage: People want to be safe. If you want people to trust the message, they need to trust the messenger.

Thank you.

Mr V. Titov joined virtually and noted that at any moment, people can find information online. It is most important to find the most appropriate communication channels and tools.

The panel discussion highlighted that good communication and effective continuous engagement build trust over a long period of time. Communications cannot just be switched on under emergency conditions. The public and the media need to be included as stakeholders in nuclear safety. The public needs to know where to go for information. To ensure effective communication, one should listen and understand different perspectives. It was mentioned that it is not always possible to ensure consistency in messages as the message changes with the stakeholder; therefore, there is a need to monitor the differences. Decisions need to be made by those in the authority but should not be made in ignorance – the public should be involved in the decision making. The point was made that expert decisions need to be communicated in a trustworthy manner to 'the interested public' and those we wish to convince, not necessarily to 'the general public'. There will always be people who disagree with people in the authority. The role of opinion makers and how to reach "beyond the converted" was discussed and it was agreed that the best way was to provide feedback and react to communication, and to engage with local journalists. Governments have a role to play in communications on nuclear matters as does the IAEA. A source should be fully trusted, as the IAEA is by newcomers, but independent sources of information also play a role. The challenge of how to engage the public positively and emotionally regarding such a grey industry as nuclear could be met by using young, enthusiastic people to give the message. Also, it is very important to ensure understanding of nuclear phraseology.

The Q&A session discussed the balance between communicating risks and benefits without the risks 'blinding' stakeholders to the positives. There was a suggestion of the creation of an "all-hazards Wikipedia" for reference as many nuclear safety authorities are not always skilled in talking to the public. It was suggested that communication is not an act to convey information but an effort to build a relationship, and there are challenges in such a polarized world to do this effectively.

5.10. PANEL 10 – FACING NEW CHALLENGES

Ms M. Crane: Panel Moderator
Mr A. Kumar Dutta: India/Nuclear Power Corporation of India
Mr A. Ferapontov: Russian Federation/Rostechnadzor
Mr D. Roberts: United States of America/Nuclear Regulatory Commission
Ms U. Stoll: Germany/Gesellschaft für Anlagen- und Reaktorsicherheit GmbH
Mr S. Zaozhan: China/Nuclear and Radiation Safety Centre
Ms M. Ziakova: Slovakia/Nuclear Regulatory Authority

The rapidly changing world inevitably reflects on the nuclear energy landscape. Considering this changing environment and new technologies, enhancing safety in the next years remains a key task for governments, regulatory bodies and operating organizations. Natural phenomena, economic trends and social expectations are difficult to foresee. New technologies impact energy demand and supply, and shape life. Events, such as pandemics, impact mobility of people and goods and affect nuclear supply chains. During this panel, participants discussed trends and global developments that have a direct impact on nuclear power in its current format. Opportunities to enhance safety under new circumstances were also discussed.

Mr A. Dutta joined virtually and noted that, from an operator's viewpoint, nuclear energy will play an important role in reducing carbon dioxide. Peer and periodic reviews, especially of preparedness against external events, are important.

Mr A. Ferapontov gave details on the recent Russian NPP programme of life extension, and nuclear fusion installations.

Mr D. Roberts noted that the world is evolving and nuclear will be impacted. Nuclear energy is a proven carbon free option. The United States has experienced 350 reactor years beyond 40-year life; however, parts obsolescence is a problem. New technologies will continue to emerge. Regulation of innovative designs will require global collaboration.

Mr U. Stoll discussed Germany's forward programme. He questioned how designed safety systems will cope with the higher external air and water temperatures due to climate change. New nuclear power, if it takes 30 years to build, will be too late for climate change. He noted that trust between all players was key during COVID-19, for example during remote inspections.

Mr S. Zaozhan joined virtually and noted that cyber safety, counterfeit and fraudulent products, and malicious attacks are challenges to the industry. Nuclear costs seem to be increasing, while other clean energy technologies' cost is decreasing. Elimination of accident sequences might lead to changes in the defence in depth philosophy.

Ms M. Ziakova stated that one needs safe, secure, reliable and climate friendly energy in the future, and that one would need to continue to be vigilant with aspects that have already been evaluated. Standardization of designs and harmonization of regulatory approach are key for future nuclear programmes.

Ms Crane summarized panellists' opening remarks as covering the challenges of climate risks, societal and economic risks, technological challenges (new and extensions), and regulatory challenges.

During their ensuing discussion, the panellists mentioned that, in order to meet these challenges, nuclear power could become self-protecting vis-à-vis climate change. They discussed the challenges climate change is placing on energy production in general and NPP operation in particular, how NPPs can adapt to the challenges, and what kind of designs can meet these challenges. They agreed that current NPPs are robust enough to meet the challenges of climate change and noted that some new designs can never reach the melting point of the fuel; however, some safety margins may change, and the operators will have to monitor these. Standardization would make nuclear more competitively priced and quicken deployment of new units. Regulation similar to that of the aviation industry might also help, as might moving away from technologies only proven for nuclear power. They highlighted that the nuclear industry removes the responsibility for safety from the equipment designer, but the aviation industry does not. It seems that levels of trust are unchanged over the past few decades and those firmly against nuclear power cannot be persuaded to be supportive of it. They also saw cyber security as one of the most important challenges to be met.

The Q&A session following the panel discussion highlighted the following for consideration: Germany still operates research reactors, so they still have a role to play. While SMRs could be the answer to the increasing costs of nuclear power development, political actions may make other renewables cheaper. Vendors are entering newcomer countries, providing power plants and then improving the safety of designs. Molten salt reactors were seen to be a good idea to deal with radioactive waste but not the best option purely for power generation. Russian Federation is planning to build four more floating reactors but with a different design compared to the current one. The audience noted that lack of qualified staff is a problem across Europe and construction of geological disposal facilities for radioactive waste is a challenge for the future. Regarding accidents, it is important to refer to other causes such as war and armed conflicts as well. Is it best to shut down NPPs in the face of extreme external events when their electricity might be useful in response to and minimisation of potential consequences ofthese events?

5.11. PANEL 11 – CALL FOR ACTIONS – MAINTAINING THE MOMENTUM

Mr M. Weightman: United Kingdom/Conference President and Chair Session A
Ms R. Velshi: Canada/Chair Session B
Mr D. Dorman: United States of America/Session C (representing Mr C. Hanson)
Mr N. Ban: Japan/Chair Session D
Mr C.-M. Larsson: Australia/Chair Session E

Ms Lydie Evrard: IAEA Deputy Director General and Head of the Department of Nuclear Safety and Security

Recognizing that discussion is not enough to defend against complacency and demonstrate the safety of the nuclear industry, this panel, chaired by the President of the Conference and consisting of the Session Chairs, used the key messages from the conference, coupled with insights gained from speakers throughout the week to identify where future efforts and momentum would be most beneficial in four main areas: Enhancing openness, better preparation for the wider use of nuclear power, embedding the lessons from the Fukushima Daiichi accident and transferring knowledge for the future. The outcomes from this panel were incorporated into the President's summary and Call to Action, a tool that the IAEA can take forward and use to help shape the future of nuclear safety in the 21st century.

The panellists noted that, to enhance openness and make national and international systems even more robust, interaction and interface need to be encouraged. We need to be better at, and have better guidance on, communicating our messages around safety, both the risks and the benefits, especially to youth, using different methods, e.g. social media. Part of this is how to better explain and communicate how safe is safe enough. EPR has to be clear and transparent and needs to include the public health community, as our response has non-radiological consequences. A more holistic approach to low dose radiation is needed to differentiate between the attribution of risk and inference of risk. In protecting people against radiation exposure, both the radiation risks and the psychosocial effects of radiation protection need to be taken into account. The radiation health effects need to be balanced with the psychosocial effects and the effect of protective actions. In recovering from a nuclear emergency, the answer to what is safe enough needs to be established in advance in the consultation phase and communicated to the public as a range of acceptable end states. We need to take into account the social aspect of improving safety, providing information so that people can make their opinions in an informed way. We need to talk and listen to local people; it is less about challenges and more about understanding and providing information, explaining what is being done and why.

In order to better prepare for the wider use of nuclear power, the panellists suggested that the global governance structure should be reviewed, to ensure that there are appropriate governmental and institutional frameworks, in accordance with and in adherence to the IAEA safety standards, sufficient availability of technical support (especially for new users) and effective and efficient international collaboration for possibly thousands of NPPs. The nuclear industry will have to consider if appropriate governance structures currently exist and how best to continue to foster a culture for safety, that also includes security, in regulators, operators and policy and decision makers, including non-nuclear people, so that we can equip them with the tools that incorporate the lessons learned from the Fukushima Daiichi NPP accident and lessen any impact to the public health system and the public itself. We need to find more effective ways to communicate with the society honestly and openly about the wider use of nuclear applications, clarify the scope and target of communication, and make use of knowledge from the social and behavioural sciences, particularly to engage youth. We need to be prepared to recover as well as respond as transitions between phases occur gradually and in different ways. There is a need for clear standards and guidelines for levels of radiation in commodities thus avoiding harm through lack of understanding of the risk, with international cooperation and commitment. Recognizing, anticipating and adapting to technological and societal changes will be important, as will knowledge management and capacity building. Both a questioning and a listening attitude will need to be incorporated.

As for embedding the lessons of the Fukushima Daiichi NPP accident, much has been done in existing nuclear installations, but it needs to be documented and explained to the public. Regulatory bodies need to verify that new technology designs have embedded the lessons learned. Both good and poor practices in off-site radiation protection and the effects of evacuations and large scale contamination need to be reviewed. The question to be asked needs to be changed from a question of doses to what is best for people's well-being. The nuclear industry can learn from other industries and natural disasters how to talk about risk so that it is better understood and communicate so that risk informed decisions can be made without fear and stigma of radiation that can lead to misinformation. The non-radiation effects of the Fukushima Daiichi NPP accident have had more impact than radiation exposure on people's health and well-being. It is not a question of improving safety at any cost but a balance to further strengthen safety where reasonable to do so. Improving safety does not necessarily mean that nuclear applications were previously unsafe; improving safety and learning are continuous processes. It is the responsibility of all stakeholders to improve this learning process, using knowledge transfer from other industries and incorporating the use of techniques such as cost–benefit analysis.

To transfer knowledge for the future, action is needed. This knowledge is hard won and open sharing of all experience, especially that of the Japanese, is essential and access needs to be universal. Japan is now the world's leading expert on nuclear accidents, and we need to ensure, given the very long timeframe and unique circumstances involved, that there is a repository of knowledge and new technology that has been developed for these particular circumstances, that is searchable and accessible for people who might need it in future, i.e. what was done, why, and how, particularly in decommissioning. There is an intergenerational aspect of knowledge transfer. The IAEA has developed tools such as peer review, advisory services and schools for capacity building and these are available for future generations. Resources are necessary for knowledge transfer. Many opportunities for harmonization of requirements and working together are available, increasing gender equity and diversity, to support embarking countries with investments in both technological and human resources. Wider interactions with more issue focused meetings and other mechanisms to encourage bilateral and regional cooperation would be beneficial. Some of our existing tools have reached maturity and there might be a need to review our legal instruments; the world is markedly different than when they were established. We need to develop advanced tools for the future. More efficient cooperation, coordination and consistency of international organizations' work on the same topics would also be helpful.

The Chair concluded the panel discussion with a round table on how best to maintain the momentum of the conference, with panellists suggesting that the work done at this conference, defined as actions items, should be presented to the IAEA for consideration and findings could be shared with other topical events for more detailed consideration. Follow up discussions at regular intervals would optimize the value of the work done for this conference.

6. SUMMARY OF SIDE EVENTS

6.1. SIDE EVENT – PERFORMING NUCLEAR SAFEGUARDS AT FUKUSHIMA

Mr J. Persin: IAEA/Department of Safeguards

Through a set of technical measures, the IAEA verifies that States are honouring their international legal obligations to use nuclear material and technology for peaceful purposes only.

To support the drawing of safeguards conclusions, the IAEA Department of Safeguards relies heavily on the timely access of inspectors to nuclear material and facilities. The conditions on the Fukushima Daiichi site after the impact of the Great Tohoku Earthquake and tsunami posed unprecedented and continuously evolving challenges in fulfilling this mandate. To overcome these challenges, the IAEA developed innovative measures that support the IAEA's mission, with close cooperation from the Japan Safeguards Office (JSGO) of the Japan Nuclear Regulation Authority and the Tokyo Electric Power Company (TEPCO). These innovations have combined technological advances and adaptations to the safeguards approach to the site.

Mr J. Persin provided a presentation on the safeguards activities undertaken by the IAEA at the site in the 10 years since the accident, from the post-accident response up to the current situation, the progress made in re-verifying the nuclear material that was left inaccessible by the accident, and the innovative technologies developed in response to the challenges on the site. Due to the emergency, all safeguards activities were suspended from 2011 to 2012. Enhanced cooperation was established with the Fukushima Task Force and this cooperation continues. Access to areas was limited due to dose rates. Monitoring systems were installed with remediation measures. Safeguards have been re-verifying the nuclear material from the spent fuel ponds since 2013. There were challenges in storage capacity, with new dry storage established. One hundred days of inspection, involving 22 different inspectors, took place in 2013/2014 alone. A variety of verification challenges, such as the fuel conditions, meant different methods had to be used, such as authorizing and using a digital Cerenkov viewing device system and innovative underwater NDA setups. They used the ALARA approach to verification measures. Next steps include a shift from a post-accident site to an industrial decommissioning one. They will be emptying the ponds and verifying nuclear material from the damaged cores and dealing with the debris, monitoring the dismantling activities and verifying material in unknown form. There has been new construction which will require its own safeguards measures. All of this activity will span decades, so organizational planning and resiliency to sustain and adapt over many years will be crucial. Procedures need to be optimized and information managed for future use. Knowledge transfer is and will be important at both the team and individual levels. Safeguards need to present a strong awareness of present activities, maintain information for future use, and anticipate the evolution of verification measures.

Answers to audience questions explained that some techniques may be reused in regular activities; additional surveys below the rubble are planned but this is not a safeguards issue, Japan will also be studying the material transferred and digital twins could be helpful, however not in the beginning, as these are qualitative measures only. The objective of core retrieval for safeguards is to assess the content of the fissile material and verify the inventory.

6.2. DIRECTOR GENERAL SPECIAL SESSION – SAFETY RELATED ASPECTS OF HANDLING ADVANCED LIQUID PROCESSING SYSTEM TREATED WATERS AT THE FUKUSHIMA DAIICHI NPP

Ms H. Vaughan Jones: Moderator
Mr R. M. Grossi: IAEA Director General
Mr K. Hagiuda: Japan/Ministry of Economy, Trade and Industry

Mr K. Yumoto: Japan/Nuclear Accident Disaster Response
Mr J. Matsumoto: Japan/Tokyo Electric Power Company
Mr N. Ban: Japan/Nuclear Regulation Authority
Mr G. Caruso: IAEA/Department of Nuclear Safety and Security

In April of 2021, the Government of Japan announced its plan to release ALPS treated water from the Fukushima Daiichi NPP into the sea through controlled discharges. The Government of Japan requested the assistance of the IAEA to ensure that the discharge takes place in a safe and transparent manner. Rafael Mariano Grossi, Director General of the IAEA, committed to support the Government of Japan, before, during and after the water discharge, and the IAEA Secretariat has started initial planning and implementation activities associated with its review. A Task Force comprised of the IAEA Secretariat, as well as international experts, has been established and will review the Government of Japan's activities related to the treated water discharge. In this Special Session, key officials from the IAEA and Japan shared information about recent progress and future activities.

Mr Grossi stated that this issue has attracted a great deal of attention; in this regard, it is important that we take stock here and share what has been learned. The Agency's role is to review Japan's policy and plans and to add value through its safety standards as a benchmark for the review of the handling of the ALPS treated waters, while demonstrating to the world that Japan values the international community. The Agency's role would not be as a bystander, but as an active and independent participant. To conduct its work, the Agency has formed an international Task Force that remains neutral, impartial, and scientifically robust. The Agency intends to share the results of its work in real-time in a transparent manner. In this regard, it is hoped that the IAEA is viewed as the nuclear scientific authority.

Mr K. Hagiuda noted that 10 years have passed since the accident; decommissioning and reconstruction continue. ALPS water is an issue that is being tackled. Japan's policy is that, given that TEPCO complies with national laws and international standards, ALPS water is to be discharged in two years. Following the discharge, monitoring will be conducted in a thorough and transparent manner, and the results of reviews will be distributed widely. Working with the IAEA's rigorous monitoring and review provides another layer of assurance that the work is safe.

Mr K. Yumoto provided the Japanese government perspective with an overview of the government policy on the handling of ALPS treated water. There have been over six years of technical discussions concerning the treatment of ALPS water with continuous stakeholder involvement and communication, including transparency with the international community. The IAEA will review work under the policy before, during, and after discharge. A website has been provided for more information.

Mr J. Matsumoto provided the operator perspective with a technical overview of the work associated with the handling of ALPS water including its removal, redirection, and preventing leakage. Treatment of the water using ALPS has reduced annual doses to below regulatory levels (1 mSv). Production of treated water has decreased significantly; however, the water cannot be stored indefinitely (there is a limit to storage space). He gave an overview of the ALPS process including the steps for treatment, measurement, analysis, and dilution prior to discharge. The water will be reviewed by a third party prior to discharge. He also outlined the plan for monitoring post discharge (water and biota) and stressed the importance of communication for fostering understanding, including the development of information in multiple languages and a website.

Mr Ban noted that, due to the accident, managing the site under normal regulations was not possible; therefore, legislation for a special arrangement for managing the facility was formulated. He provided an overview of the ALPS process, criteria, and environmental monitoring from the regulatory perspective, with oversight of the revision of TEPCO's implementation plan for ALPS. Review meetings for the revision of the implementation plan are open to the public and broadcasted via YouTube and will be reviewed by the IAEA.

Mr Caruso explained that the Government of Japan requested the IAEA to review the implementation plan for the discharge of ALPS treated water using its safety standards. Reviews will be conducted before, during, and after the planned discharge to ensure safety and transparency. The focus of the Agency's work in the short term will be to answer two questions: Are the plans of the operator in compliance with the Government of Japan's expectations/policy? Is the regulatory body's oversight in compliance with the Government of Japan's expectations/policy? The Agency will use its safety standards as criteria and has created a diverse Task Force with representation from 11 countries. The Secretariat will be the leading component of the Task Force, while the international experts will serve in an advisory capacity. The components of the IAEA review are safety assessment, regulatory activities, and sampling and environmental monitoring (including corroboration of key data). Outputs of the review will include interim progress reports and updates to the public as well as the international community at meetings, conferences and briefings to Member States. The timeline includes review missions on safety assessment in 2021, regulatory review and corroboration of data in 2022 with activities to continue through 2023.

During the following Q&A session, it was suggested that monitoring of contamination before and after treatment be done by the IAEA with all data publicly provided. Panellists noted that all relevant interventions will be taken under advisement and the situation will continue to be monitored in accordance with stated plans. Regarding a communication plan, Mr Grossi stated that communication will be an ongoing process, avoiding periods of silence, and the IAEA is working with IT to develop innovative platforms for engagement. Mr Matsumoto noted that TEPCO is preparing to communicate as broadly as possible to as wide an audience as possible. In order to minimize bias of the environmental assessment models to avoid over or under estimation, the NRA will review only the outputs of the assessment and the IAEA will review the evaluation process. Daily measurement of seven major nuclides will verify the performance of the ALPS system. The plan for the waste of the ion exchange will be completed under the oversight of the NRA.

7. CLOSING SESSION

Mr M. Weightman: Conference President
Mr G. Caruso: Scientific Secretary
Mr R. M. Grossi: IAEA Director General

In the closing session, the President of the Conference summarized his findings and conclusions from the conference sessions and panels and proposed a 'President's Call to Action'.

7.1. PRESIDENT OF THE CONFERENCE – CONCLUSIONS OF THE CONFERENCE

A copy of his closing presentation has been provided as follows:

Director General, honoured guests, delegates I now give you my thoughts on this most important, timely and seminal conference.

But, first let me thank you all for your expert and very valuable contributions to this conference: speakers, chairs, panellists and delegates. Let me also thank the Director General for his excellent idea of staging such a conference with a different format, different expectations and different outcomes; Gustavo for being such an excellent Scientific Secretary; and especially all the IAEA staff and others who have worked tirelessly for us all.

I now will provide from what I have heard and seen a summary of my observations, and most importantly, a Call for Action. In doing so I have to bear in mind the overall goal: "Safe nuclear power for all" as part of providing secure clean energy for humankind and hence part of the solution to climate change. But this conference is only about the safety part of achieving that goal. Let me provide a little context.

Now, my thoughts on this conference. There are many. I will summarize them in my report. This comes in a couple of weeks' time. You can read that at your leisure. But for now, I share with you the headlines.

First though, I want to mention the three special events. They were excellent and most informative. I found most interesting the Youth Panel which filled me with much hope that the future is in safe hands.

My other main thoughts are set out in line with the 'Enabler' themes we have discussed at this conference. These are needed to secure our Goal. There are other enablers, particularly around financing mechanisms and project delivery, but these are for another time and place. We have heard some great presentations and speeches, and the discussions in the panels and through the Q&A have been most stimulating.

The First Enabler then:

Greater International Collaboration

To achieve "Safe Nuclear Power for all" requires us all to make tremendous efforts, together. We have heard how much our Japanese colleagues have achieved in the last ten years at Fukushima Daiichi and in the surrounding area through their dedication, expertise and by working together with a common goal assisted greatly by the international nuclear community. An example to us all. Also, their openness and transparency has enabled the international community to learn lessons and improve nuclear safety, with greater understanding. We heard from the leaders of various international bodies about their excellent work and their thoughts for the future. We need to harness this tremendous capability in a more collaborative way if we are to achieve our goal. This should underpin all that we do.

So how can we do that? Reflecting on the presentations of the various international institutions and those from various countries, the excellent work they have done and are doing, as a first step we should bring together all the work of these and other institutions into a common overview of existing lessons learned, recommendations and actions outstanding to demonstrate what has been achieved, and show that the learning continues in a more collaborative way as we help each complete the work together.

Of course, fulfilling such an action will embrace, in general terms, several of the enablers.

But there are some aspects that I believe warrant particular attention. I cover these under a heading of:

Busting the myth

One of the key outstanding issues that we need to address, if we are to stop the confusion we sow in people's minds and be able to earn their trust, is that of low doses of ionizing radiation and how we then go on to ascribe thousands of deaths to an accumulation of individual trivial doses. People are exposed to risks all the time from a wide range of sources. Some they do not accept as being too high, some they tolerate because they see the benefits outweigh the detriments, and some they perceive as too trivial to worry about (notice) and ignore. COVID-19 has taught us that people can understand that we have to balance risks and are better able to put them into perspective. We have to build on this and address those aspects under our control that sow unnecessary confusion and anxiety.

Optimum decisions

One related aspect we heard about during this past week, along with many other excellent contributions, was the lesson learned about decision making for evacuation and recovery operations. Decisions setting limits so low that reasonable actions cannot be taken or do not reflect reality or balance the risks and benefits or are different in different jurisdictions when commodities are transported worldwide, serve no one and only lead to mistrust of authorities, unnecessary alarm and sub-optimum results. The international collaborative approach I advocate above should address these issues as well.

Safe enough?

I have heard this week about the many lessons that have been learned and acted on to improve safety. I have also heard about the new developments and challenges. And, of course, this conference is about this progress after the Fukushima Daiichi NPP accident and about building on the lessons learned to further strengthen nuclear safety with a changing environment. But it is also about balance. You do not change something just because you can or just because you have a new design that may be demonstrated theoretically to be safer. You have to make a judgement on the balance of decrease in risk against the detriments of doing so – this is what I spent my life doing as a nuclear regulator where my duty was to make decisions on what was reasonably practicable (in UK parlance). Nuclear is safe by any common comparison but that does not stop us from seeking reasonably practical improvements. It is what keeps us on our toes.

Remember though what Confucius said: "To go beyond is as wrong as to fall short."

Inclusive leadership – a keystone

All this requires great effort but led in a way that generates involvement of heart as well as mind, commitment, excitement and utilizing the attributes of all. Humble leadership. We as leaders have much to learn from others – you only had to listen to the Special Youth Panel. It is what holds us all together.

Again, this is an enabler that would greatly benefit from working together as leadership requirements penetrate through all areas of the nuclear enterprise. Both the IAEA and WANO have done much in this area.

International legal instruments – A foundation stone

To drive forward enhanced international cooperation and collaboration, and further strengthen nuclear safety worldwide, we need a firm basis to do so. That is provided by international legal instruments. From our panel discussion, I conclude that we do have a firm basis, but it can be reinforced to be more effective in the years to come to better face the changing challenges, especially to facilitate a wider safe use of nuclear power.

We need to be changing as our environment changes as it has and will rapidly. Remember Darwinian theory: "Adapt to your changing environment or you die out."

Earning trust

To reach the goal, we have to earn the trust of people. We have made a start this week and demonstrated to our peers that we have learned the lessons and look forward to making further improvements, addressing he challenges where reasonably practicable to do so. But this will not be enough to earn the trust of people, policy makers and politicians.

It will be a long and difficult journey but one we have to travel. One that can be done.

Society is changing. It is more pluralistic and interconnected globally. People have many routes to accessing information. Some of it is misleading. Remember Darwin again.

I heard how people have learned from the response to the Fukushima Daiichi nuclear accident and are taking actions to be more open. And I witnessed this for myself earlier last week at the NDF International Forum in Fukushima province how engaging with local people over the last years, providing information, listening to them and responding to their concerns can change the environment in which you work – from confrontation and fear to one of tolerance, understanding and growing trust.

So in conclusion, this enabler is essential, will be difficult but attainable. And it came up many times during the week and one that is emphasized by a simple model set out in INSAG-27, Ensuring Robust National Nuclear Safety Systems – Institutional Strength in Depth.

As in any system, effective interfaces are key – without them, the system is broken. Fukushima Daiichi emphasized that. I therefore agree that enhanced openness is essential to achieving the goal.

Challenges

We heard there are many. But I also heard commitment, energy and capability to face up to them and overcome them. But it will need effort, great effort, combined effort, from all, and we need to continue to learn from each other and others outside the nuclear community. There will also be opportunities. We need to be prepared to grasp them, holding them high so they shine a light for others to follow.

So, challenges will be many, but together we can meet them ("Yes we can!") And take the opportunities that sit alongside them. To do so, we need to be responsive, agile and combine our efforts, not afraid to use the best athlete.

Action is needed

Talk and words may be fine, but they are not enough. Action is needed. We will have failed if we do not take action. The world will not be able to make the fullest use of safe nuclear power. People around the world will not have full access to the clean energy they need to live healthy, satisfying lives in peace and harmony. So, I am calling for action based on your inputs and discussions. It will be detailed later but for now, I summarize my thoughts based on what I have heard and seen at this conference. It will be expanded, and I hope taken by the Secretariat.

There are four areas where I call for action: Enhance Openness; Safety Commitment for Wider Use of Nuclear Power; Embed Lessons from Fukushima Daiichi; and Transfer the Knowledge for the Future. Under these, I have identified the following:

1) Enhance Openness

- CA.1: Summarize all the work in response to the accident as a basis for the demonstration of the International Institutions working together, lessons learned globally and improvements made over the last ten years; and, as a basis for more effective working together in the future.

- CA.2: Identify and implement an effective mechanism to review annually the Action Plan and its progress, such as a side event at the IAEA General Conference.

- CA.3: To help enable better public understanding, review of present system of radiation protection with regard to the inference of risk from low doses, and the use of dose limits when there are different balances to be made.

- CA.4: Extend guidance, including practical examples in different circumstances, on decision making when balancing risks associated with ionizing radiation vs other risks or benefits.

- CA.5: Conduct research into human behaviour, risk perception and risk communication, to develop guidance and tools for earning the trust of the public and stakeholders in the context of a global strategy.

2) Safety Commitment for Wider Use of Nuclear Power

- CA.6: Review together and enhance the global system (governance framework, technical support etc.) for nuclear safety to ensure full cycle oversight in light of the potential need to significantly increase the number of nuclear reactors, other nuclear installations, radioactive waste facilities and the host nations.

- CA.7: Take stock of the international experience of recovery from major nuclear accidents to date, including the direct and indirect impact on health and well-being of actions taken during both the emergency and recovery, consolidate the observations and review their implications for the IAEA safety standards and decision guidelines.

- CA.8: Expedite the work on unified standards and guidelines for radiation levels of commodities (food and others) that are based on risk, and communicate these including the risk assessment.

- CA.9: Include generic planning for recovery in the plans for new facilities, establish the division of responsibilities between different parties, and include arrangements for recovery after an accident in the consultation process preceding the authorization of new facilities.

- CA.10: Promote the concept of "cultures for safety", making use of individual national cultures.

3) Embed Lessons from Fukushima Daiichi NPP accident

- CA.12: Invite the Contracting Parties to the CNS at the 9th Review Meeting in 2023 to focus attention to the timely completion of remaining safety improvements at existing NPPs resulting from lessons from the Fukushima Daiichi accident.

- CA.13: Ensure that standards for new reactor technologies include principles that appropriately take into account the lessons from the Fukushima Daiichi accident.

- CA.14: Undertake a review to determine whether a comprehensive view of public health consequences of the accident (that is, radiological risk averted by evacuations vs the non-radiological health risks demonstrated from (near term) evacuations and (long term) displacement of the population) suggests changes to protective action guidelines for decision makers following an accident.

4) Transfer of Knowledge for the Future

- CA.15: From the Fukushima Daiichi NPP accident, experiences compile a knowledge management document to guide planners in the issues they will need to consider and address when responding to and recovering from a nuclear accident.

- CA.16: Provide a forum for sharing and promoting best practices in the remediation of decommissioned NPPs.

- CA.17: Develop a knowledge management record of the challenges and best practices arising from the Japanese experience of decommissioning and decontamination.

No doubt as we reflect further on this conference there may be more, or they may be changed.

Am I now content that this conference of which I am honoured to take part in is a success? Partially. I am content with the discussions, conclusions and the Call for Action but again the people we serve might rightly say these are just words. We need to see that the promises, the actions, are in fact delivered. So how do we do that? We need a concrete mechanism to make it real. We need commitment of all to deliver it. We need a new mechanism to ensure it is delivered visibly, transparently for all.

I propose, Director General, that a commitment is made to develop and publish a concrete Call for Action based on the deliberations of this conference and that a mechanism is devised to review and progress it, perhaps a side event at each General Conference of the IAEA. Such an event would be open and a joint event with other international institutions. I truly hope that this proposal will be taken up. And I ask all to be fully committed to delivering such a Plan for Action with all vigour together.

Then I will be content that this conference has been a great success. And, in years to come it be seen as making the difference, be seen as an important legacy of your reign as Director General, but most of all provides a basis over the next few years for achieving the goal of safe nuclear power for all, being part of the solution to our climate change crisis and fulfilling our duty to the people we serve.

Director General, honoured guests, chairs, delegates I commend the outcome of this conference to you, and I thank you.

7.2. CLOSING REMARKS

7.2.1. Scientific Secretary of the Conference – Closing Remarks

Mr G. Caruso's closing remarks as the Scientific Secretary for the conference have been summarized as follows:

Mr Caruso noted that there were 689 participants, 131 speakers and panellists, 216 observers, 68 countries and 7 international organizations involved in the conference. Feedback over the week indicated that this had been a very valuable conference, and he thanked everyone involved in preparing for and implementing the conference.

7.2.2. IAEA Director General – Closing Remarks

The Director General's closing remarks have been summarized as follows:

Director General Grossi thanked Mr Weightman for his service to the nuclear community. He noted that this conference was the right thing to do; it was not an autopsy of the accident, nor an exercise in regret and regurgitation. The community took stock and made progress in a shift that is embodied in the four areas noted by Mr Weightman in his address. He encouraged conference participants to persevere in the effort to better inform stakeholders in a timely manner and to help the IAEA in its aims for better communication. He reflected that the Fukushima Daiichi NPP accident was seen as the swan song of nuclear, however, safe nuclear power for everyone is part of the solution to climate change and this is the resurgent image of the nuclear industry. Safe nuclear power exists in a broad context, and need to consider matters of public trust need to be considered, as well as climate change and the involvement of future generations in the global development and success of nuclear power. Knowledgehasbeen built on the lessons learned from the last ten years, and together one needs to be preparing for the next decade. This week has been re-energizing, and with the Call for Action, we have a guide for the work to be done in the next ten years. His final message was to stay tuned for the changing world

ANNEX I. PROGRAMME OF THE CONFERENCE

MONDAY, 8 NOVEMBER 2021

09:00-10:00   OPENING SESSION   **Boardroom B/M1**

Time	Name	Designating Member State/Organization
	Naga Munchetty	*Moderator*
09:00–9:10	**Rafael Mariano Grossi**	IAEA Director General
9:10–9:20	**Takeshi Hikihara**	Ambassador of Japan
9:20–9.30	**Mike Weightman**	Conference President (United Kingdom)
9:30–9:40	**Gustavo Caruso**	Scientific Secretary

Keynote speakers

Time	Name	Designating Member State/Organization
9:40–9:50	**Lydie Evrard**	IAEA Deputy Director General and Head of the Department of Nuclear Safety and Security
9:50–10:00	**Hajimu Yamana**	Japan Nuclear Damage Compensation and Decommissioning Facilitation Corporation

PART I – INTERNATIONAL ORGANIZATIONS PERSPECTIVE

**10:00-12:00   SESSION   A   –   CONTRIBUTION   OF   Board Room B/M1
INTERNATIONAL ORGANIZATIONS TO GLOBAL
SAFETY**

Time	Name	Designating Member State/Organization
	Mike Weightman	*Session Chairperson* *United Kingdom/Conference President*

Time	Name	Designating Member State/Organization
10:00–10:12	**William Magwood**	Organisation for Economic Co-operation and Development/Nuclear Energy Agency (OECD/NEA)
10:12–10:24	**Ingemar Engkvist**	World Association of Nuclear Operators (WANO)
10:24–10:36	**Gustavo Caruso**	IAEA/Department of Nuclear Safety and Security
10:36–10:48	**Gillian Hirth**	United Nations Scientific Committee on the Effects of Atomic Radiation (UNSCEAR)
10:48–11:00	**Gerhard Graham**	Preparatory Commission for the Comprehensive Nuclear-Test-Ban Treaty Organization (CTBTO)
11:00–11:12	**Gerd Dercon**	Food and Agriculture Organization of the United Nations (FAO)
	Carl Blackburn	Food and Agriculture Organization of the United Nations (FAO)
11:12–11:24	**Joaquim Pintado Nunes**	International Labour Organization (ILO)
11:22–11:36	**Lars Peter Riishojgaard**	World Meteorological Organization (WMO)
11:34–11:48	**Maria Neira**	World Health Organization (WHO)
11:48–12:00	**Mike Weightman**	Session wrap-up

PART II – LEARNING LESSONS

**14:00-15:30   SESSION B – ENSURING THE SAFETY OF   Board Room B/M1
NUCLEAR INSTALLATIONS**

Time	Name	Designating Member State/Organization
	Rumina Velshi	*Session Chairperson* *Canada/Canadian Nuclear Safety Commission*
14:00–14:10	**Keiichi Watanabe**	Japan/Nuclear Regulation Authority
14:10–14:20	**Fedor Aparkin**	Russian Federation/ROSATOM State Atomic Energy Corporation
14:20–14:30	**Petteri Tippana**	Finland/Radiation and Nuclear Safety Authority
14:30–14:40	**Anne Pelle**	France/Électricité de France
14:40-15:30	**Q&A**	

16:00-17:30 **PANEL 1 – ENSURING THE SAFETY OF NUCLEAR INSTALLATIONS – MINIMIZING THE POSSIBILITY OF SERIOUS OFF-SITE RADIOACTIVE RELEASES** **Board Room B/M1**

Name	Designating Member State/Organization
Naga Munchetty	*Panel Moderator*
Mark Foy	United Kingdom/Office for Nuclear Regulation
Michael Franovich	United States of America/Nuclear Regulatory Commission
Jinho Lee	Republic of Korea/Korea Institute of Nuclear Safety
Jean Christophe Niel	France/Institut de Radioprotection et de Sûreté Nucléaire
Rosa Sardella	Switzerland/Swiss Federal Nuclear Safety Inspectorate

TUESDAY, 9 NOVEMBER 2021

**09:00-10:30   SESSION C – PREPARING AND RESPONDING   Board Room B/M1
TO A POTENTIAL NUCLEAR EMERGENCY**

Time	Name	Designating Member State/Organization
	Christopher Hanson	*Session Chairperson* *United States of America/Nuclear Regulatory Commission*
9:00–9:10	**Tomohiko Makino**	Japan/Nuclear Disaster Management Bureau
9:10–9:20	**Catarina Danestig Sjogren**	Sweden/Swedish Radiation Safety Authority
9:20–9:30	**Hessa Almarzooqi**	United Arab Emirates/Federal Authority for Nuclear Regulation
9:30–9:40	**Tasos Zodiates**	International Labour Organization
9:40–09:50	**Marcus Grzechnik**	Australia/Australian Radiation Protection and Nuclear Safety Agency
09:50--10:30	**Q&A**	
10:30–11:00	*Coffee/Tea Break*	

**11:00-12:30   PANEL 2 – PREPARING AND RESPONDING   Board Room B/M1
TO A POTENTIAL NUCLEAR EMERGENCY—
ROBUST PREPAREDNESS ARRANGEMENTS**

Name	Designating Member State/Organization
Naga Munchetty	*Panel Moderator*
Kajal Kumar De	India/Nuclear Power Corporation of India
Tomohiko Makino	Japan/Nuclear Disaster Management Bureau
Susan Perkins	United States of America/Nuclear Energy Institute

Name	Designating Member State/Organization
Patricia Wieland	Brazil/Brazilian Association for the Nuclear Activities Development

14:00-15:30 SESSION D – PROTECTING PEOPLE Board Room B/M1
AGAINST RADIATION EXPOSURE

Time	Name	Designating Member State/Organization
	Nobuhiko Ban	*Session Chairperson* *Japan/Nuclear Regulation Authority*
14:00–14:10	**Abel Gonzalez**	Argentina/Argentine Regulatory Authority
14:10–14:20	**Evgeny Metlyaev**	Russian Federation/The Federal Medical and Biological Agency
14:20–14:30	**Todd Smith**	United States of America/Nuclear Regulatory Commission
14:30–14:40	**Gerry Thomas**	United Kingdom/Imperial College London
14:40–14:50	**Zhanat Carr**	World Health Organization
14:50–15:00	**Jacqueline Garnier-Laplace**	Organisation for Economic Co-operation and Development/Nuclear Energy Agency
15:00-15:30	**Q&A**	

16:00-17:30 PANEL 3 – PROTECTING PEOPLE AGAINST
RADIATION EXPOSURE—ATTRIBUTING Boardroom B/M1
HEALTH EFFECTS TO IONIZING
RADIATION EXPOSURE AND INFERRING
RISKS

Name	Designating Member State/Organization
Melinda Crane	*Panel Moderator*
Trevor Boal	Australia
Christopher Clement	International Commission on Radiological Protection
Michiaki Kai	Japan/Nippon Bunri University
Wolfgang Mueller	Germany/German Commission on Radiological Protection
Masaharu Tsubokura	Japan/Fukushima Medical University
Shang Zhaorong	China/Nuclear and Radiation Safety Centre

WEDNESDAY, 10 NOVEMBER 2021

09:00-10:30   SESSION E – RECOVERING FROM A   Board Room B/M1 NUCLEAR EMERGENCY

Time	Name	Designating Member State/Organization
	Carl-Magnus Larsson	*Session Chairperson* *Australia/Australian Radiation Protection and Nuclear Safety Agency*
9:00–9:10	**Akira Ono**	Japan/Tokyo Electric Power Company
9:10–9:20	**Keiichi Yumoto**	Japan/Ministry of Economy, Trade and Industry
9:20–9:30	**Tatsuro Sagawa**	Japan/Ministry of the Environment
9:30–9:40	**Masaharu Tsubokura**	Japan/Fukushima Medical University
9:40–9:50	**Oleksandr Novikov**	Ukraine/Special State Enterprise Chornobyl Nuclear Power Plant
9:50–10:00	**Analia Canoba**	Argentina/Argentine Regulatory Authority

Time	Name	Designating Member State/Organization
10:00-10:30	**Q&A**	
10:30–11:00	*Coffee/Tea Break*	

11:00-12:30   PANEL 4 – INTERNATIONAL COOPERATION   Board Room B/M1

Name	Designating Member State/Organization
Hannah Vaughan Jones	*Panel Moderator*
Alfredo De Los Reyes	Spain/Nuclear Safety Council
Massimo Garribba	European Union/European Commission
Christine Georges	France/French Alternative Energies and Atomic Energy Commission
Olga Lugovskaya	Belarus/Gosatomnadzor
Khammar Mrabit	Morocco/Agence Marocaine de Sûreté et de Sécurité Nucléaires et Radiologiques
Suchin Udomsomporn	Thailand/Office of Atoms for Peace
Rebecca Weston	United Kingdom/Sellafield Ltd

**13:30-14:30   SIDE EVENT – PERFORMING NUCLEAR   Board Room B/M1
SAFEGUARDS AT FUKUSHIMA**

14:30–15:00 Coffee/Tea Break

15:00- **17:00**	**DG SPECIAL SESSION – SAFETY RELATED** **ASPECTS OF HANDLING ALPS TREATED** **WATERS AT THE FUKUSHIMA DAIICHI NPP**	

Time	Name	Designating Member State/Organization
	Hannah Vaughan Jones	*Panel Moderator*
15:00- 15:15	**Rafael Mariano Grossi**	IAEA Director General
15:15- 15:18	**Koichi Hagiuda**	Japan/Ministry of Economy, Trade and Industry
15:18- 15:30	**Keiichi Yumoto**	Japan/Nuclear Accident Disaster Response
15:30- 15:50	**Junichi Matsumoto**	Japan/Tokyo Electric Power Company
15:50- 16:10	**Nobuhiko Ban**	Japan/Nuclear Regulation Authority
16:10- 16:40	**Gustavo Caruso**	IAEA/Department of Nuclear Safety and Security
16:40- 17:00	**Q&A**	

Board Room B/M1

17:00-18:30   PANEL 5 – SPECIAL YOUTH PANEL—YOUTH
AND THE NUCLEAR INDUSTRY

Name	Designating Member State/Organization
Rafael Mariano Grossi	IAEA Director General

Name	Designating Member State/Organization
Ilieva Illizastigui Arisso	*Panel Moderator*
Hayden Rogers Page	*Panel Moderator*
Daiane Dantas Sardinha	Brazil
Kate Graham-Shaw	United Kingdom
Doovaraha Maheswarasarma	Sri Lanka
Ilia Meniailo	Russian Federation
Natchapon Promprasert	Thailand
Travis Scott Smith	United States of America

THURSDAY, 11 NOVEMBER 2021

PART III – PATH FORWARD

09:00-10:30 PANEL 6 – SAFETY FOR NUCLEAR DEVELOPMENT Board Room B/M1

Name	Designating Member State/Organization
Melinda Crane	*Panel Moderator*
Alexander Bolgarov	Russian Federation/ROSATOM State Atomic Energy Corporation
Fred Dermarkar	Canada/CANDU Owners Group
Satyajit Ghose	Bangladesh/Bangladesh Atomic Energy Regulatory Authority

Name	Designating Member State/Organization
Lei Ma	China/National Nuclear Safety Administration
Lukasz Mlynarkiewicz	Poland/National Atomic Energy Agency
Samy Shaaban Ata-Allah Soliman	Egypt/Egyptian and Radiological Regulatory Authority
10:30–11:00	*Coffee/Tea Break*

11:00-12:30 PANEL 7 – BUILDING INCLUSIVE SAFETY Board Room B/M1 LEADERSHIP

Name	Designating Member State/Organization
Hannah Vaughan Jones	*Panel Moderator*
Grote Gudela	Switzerland/Eidgenössische Technische Hochschule Zürich
Maria Lacal	United States of America/Palo Verde Generating Station for Arizona Public Service Company
Naveed Maqbul	Pakistan/Pakistan Nuclear Regulatory Authority
Elvira Romera	Spain/Nuclear Safety Council
Christer Viktorsson	United Arab Emirates/Federal Authority for Nuclear Regulation
Bohdan Zronek	Czech Republic/ČEZ Group

14:00-15:30 PANEL 8 – INTERNATIONAL LEGAL Boardroom B/M1 INSTRUMENTS

Name	Designating Member State/Organization
Melinda Crane	*Panel Moderator*
Dan Dorman	United States of America/Nuclear Regulatory Commission
Dana Drabova	Czech Republic/State Office for Nuclear Safety
Naoto Ichii	Japan/Nuclear Regulation Authority
Ramzi Jammal	Canada/Canadian Nuclear Safety Commission
Annatina Müller-Germanà	Switzerland/Swiss Federal Nuclear Safety Inspectorate
Bismark Tyobeka	South Africa/National Nuclear Regulator

16:00-17:30 PANEL 9 – COMMUNICATION, ENGAGEMENT, AND TRUST BUILDING Board Room B/M1

Name	Designating Member State/Organization
Hannah Vaughan Jones	*Panel Moderator*
Sama Bilbao y Leon	World Nuclear Association
Penelope Harvey	United Kingdom/University of Manchester
Jean-Luc Lachaume	France/Nuclear Safety Authority
Pilar Lucio	Spain/Nuclear Safety Council
Liane Sauer	Canada/Canadian Nuclear Safety Commission
Vadim Titov	Russian Federation/ROSATOM State Atomic Energy Corporation

FRIDAY, 12 NOVEMBER 2021

09:00-10:30   PANEL 10 – FACING NEW CHALLENGES   Board Room B/M1

Name	Designating Member State/Organization
Melinda Crane	*Panel Moderator*
Asok Kumar Dutta	India/Nuclear Power Corporation of India
Alexey Ferapontov	Russian Federation/Rostechnadzor
Darrell Roberts	United States of America/Nuclear Regulatory Commission
Uwe Stoll	Germany/Gesellschaft für Anlagen- und Reaktorsicherheit GmbH
Sun Zaozhan	China/Nuclear and Radiation Safety Centre
Marta Ziakova	Slovakia/Nuclear Regulatory Authority

11:00-12:30   PANEL 11 – CALL FOR ACTIONS — Board Room B/M1 MAINTAINING THE MOMENTUM

Name	Designating Member State/Organization
Mike Weightman	United Kingdom/Conference President and Chair Session A
Rumina Velshi	Canada/Chair Session B
Dan Dorman	United States of America/Session C
Nobuhiko Ban	Japan/Chair Session D
Carl-Magnus Larsson	Australia/Chair Session E
Lydie Evrard	IAEA Deputy Director General and Head of the Department of Nuclear Safety and Security

	Name	Designating Member State/Organization
12:30– 14:00	*Lunch Break*	

14:00-15:00   CLOSING SESSION   **Board Room B/M1**

	Name	Designating Member State/Organization
	Mike Weightman	Conference President
	Gustavo Caruso	Scientific Secretary
	Rafael Mariano Grossi	IAEA Director General

—

ANNEX II. LIST OF PARTICIPANTS

Name of Participant	Country	Role
Dif, Brahim	Algeria	Participant
Campos, Juan Martin	Argentina	Participant
Canoba, Analia	Argentina	Speaker
Godoy, Antonio	Argentina	Participant
Gonzalez, Abel	Argentina	Speaker
Politi, Adriana	Argentina	Participant
Varea, Lucrecia	Argentina	Participant
Gevorgyan, Tigran	Armenia	Participant
Manukyan, Davit	Armenia	Participant
Boal, Trevor	Australia	Speaker
Grzechnik, Marcus	Australia	Speaker
Hirth, Gillian	Australia	Speaker
Larsson, Carl-Magnus	Australia	Speaker
Maharaj, Prashant	Australia	Participant
Heitsch, Matthias	Austria	Participant
Roth, Dietmar	Austria	Participant
Akter, Sanjida	Bangladesh	Participant
Ghose, Satyajit	Bangladesh	Speaker
Lugovskaya, Olga M.	Belarus	Speaker
Wertelaers, Anna	Belgium	Participant
Calasich, Fabio	Bolivia	Participant
Almeida, Claudio	Brazil	Participant
Aguiar, Alexandre	Brazil	Participant
Dantas, Daiane	Brazil	Speaker
Jefferson Borges, Araujo	Brazil	Participant
Lima, Marcos	Brazil	Participant
Oliveira, Jair	Brazil	Participant
Pinto, João	Brazil	Participant
Wieland, Patricia	Brazil	Speaker
Al Nasser, Ahmad	Canada	Participant
Barbera, Lidia	Canada	Participant
Bilodeau, Alexandre	Canada	Participant
Clement, Christopher H	Canada	Speaker
Cole, Christopher	Canada	Participant
Dermarkar, Farid	Canada	Speaker
Fujita, Hiroki	Canada	Participant
Hartery, Lynn	Canada	Participant
Jammal, Ramzi	Canada	Speaker
Lulashnyk, Troy	Canada	Participant
Nsengiyumva, Dominique	Canada	Participant

Saleh, Mohamad	Canada	Participant
Sauer, Liane	Canada	Speaker
Shah, Manit	Canada	Participant
Tabra, Mirela	Canada	Participant
Velshi, Rumina	Canada	Speaker
Wright, Peter	Canada	Participant
Hao, Xiaofeng	China	Participant
Lei, Ma	China	Speaker
Liu, Lu	China	Participant
Sun, Zaozhan	China	Speaker
Shang, Zhaorong	China	Speaker
Yuan, Long	China	Participant
Apenanga, Romely	Congo	Participant
Franck Davy Rolland, Pina-Silas	Congo	Participant
Ilizastigui Arisso, Ilieva	Cuba	Moderator
Drabova, Dana	Czech Republic	Speaker
Honcarenko, Radim	Czech Republic	Participant
Kaderabek, Tomas	Czech Republic	Participant
Petrova, Karla	Czech Republic	Participant
Reznik, Vladivoj	Czech Republic	Participant
Svoboda, Karel	Czech Republic	Participant
Tisoň, Josef	Czech Republic	Participant
Zronek, Bohdan	Czech Republic	Speaker
Špaček, Jan	Czech Republic	Participant
Eldesouky Ibrahim, Wael	Egypt	Participant
Elmesawy, Mohamed	Egypt	Participant
Soliman, Samy	Egypt	Speaker
Tippana, Petteri	Finland	Speaker
Belgacem, Rayane	France	Participant
Gavart, Raphaël Nicolas	France	Participant
Georges, Christine	France	Speaker
Regaldo, Jacques	France	Participant
Journeau, Christophe	France	Participant
Lachaume, Jean-Luc	France	Speaker
Le Fessant, Elouan	France	Participant
Niel, Jean-Christophe	France	Speaker
Pelle, Anne	France	Speaker
Raymond, Maxime	France	Participant
De Buchere De L'Epinois, Bertrand	France	Participant
Distler, Pascal	Germany	Participant
Fessler, Andreas	Germany	Participant
Koehler, Ralf	Germany	Participant
Muench, Ina	Germany	Participant
Moshövel, Marcus	Germany	Participant

Müller, Wolfgang-Ulrich	Germany	Speaker
Segref, Daniel	Germany	Participant
Stoll, Uwe	Germany	Speaker
Ioannidou, Alexandra	Greece	Participant
Plastino, Wolfango	Holy See	Participant
Elter, Jozsef	Hungary	Participant
Mészáros, István	Hungary	Participant
Répánszki, Réka	Hungary	Participant
Magnusson, Sigurdur	Iceland	Participant
Dutta, Bijan Kumar	India	Participant
De, Kajal Kumar	India	Speaker
Dutta, Asok Kumar	India	Speaker
Hajela, Sameer	India	Participant
Thakur, Divya	India	Participant
Heryanto, Toto	Indonesia	Participant
Rahayu, Dwi	Indonesia	Participant
Sihana, Sihana	Indonesia	Participant
Sriwa, Muh Suhalmin	Indonesia	Participant
Suseno, Heny	Indonesia	Participant
Jafarian, Reza	Iran, Islamic Republic of	Participant
Sepanloo, Kamran	Iran, Islamic Republic of	Participant
Vosoughi, Amir Hossein	Iran, Islamic Republic of	Participant
Al-Nasri, Salam	Iraq	Participant
Al-Lami, Aqeel Maryoosh Jary	Iraq	Participant
Cunniffe, Anna	Ireland	Participant
Ryan, Robert	Ireland	Participant
Haquin Gerade, Gustavo	Israel	Participant
Bersano, Andrea	Italy	Participant
Monti, Sergio	Italy	Participant
Aoyama, Yoshiko	Japan	Participant
Arafune, Shuichiro	Japan	Participant
Asanuma, Ai	Japan	Participant
Aoyagi, Asako	Japan	Participant
Baba, Yasuhiro	Japan	Participant
Ban, Nobuhiko	Japan	Speaker
Goto, Yusuke	Japan	Participant
Harimoto, Yukiko	Japan	Participant
Hikihara, Takeshi	Japan	Speaker
Hiruta, Kazuhiko	Japan	Participant
Hisamichi, Nanako	Japan	Participant
Hokugo, Taro	Japan	Participant
Homma, Toshimitsu	Japan	Participant
Honzawa, Yuko	Japan	Participant
Hagiuda, Koichi	Japan	Speaker

Inoue, Naoko	Japan	Participant
Ishii, Katsuyuki	Japan	Participant
Iyonaga, Ayaka	Japan	Participant
Ichii, Naoto	Japan	Speaker
Kaneko, Shuichi	Japan	Participant
Katano, Takayuki	Japan	Participant
Kawamura, Hana	Japan	Participant
Kimura, Hitomi	Japan	Participant
Kinjo, Shinji	Japan	Participant
Kuroda, Hiroyuki	Japan	Participant
Kuwabara, Atsushi	Japan	Participant
Kai, Michiaki	Japan	Speaker
Kawamata, Kotaro	Japan	Participant
Matsune, Aya	Japan	Participant
Michikawa, Yuichi	Japan	Participant
Mitsuhashi, Yasuyuki	Japan	Participant
Mizuno, Toshiaki	Japan	Participant
Murayama, Ryosuke	Japan	Participant
Makino, Tomohiko	Japan	Speaker
Matsumoto, Junichi	Japan	Speaker
Nagai, Ryo	Japan	Participant
Nakagawa, Masaki	Japan	Participant
Niioka, Terumasa	Japan	Participant
Noda, Tomoki	Japan	Participant
Nagayoshi, Shoichi	Japan	Participant
Noda, Tomoki	Japan	Participant
Ogino, Haruyuki	Japan	Participant
Onoyama, Kai	Japan	Participant
Ono, Akira	Japan	Speaker
Sagawa, Tatsuro	Japan	Speaker
Sato, Gyo	Japan	Participant
Sato, Kazuko	Japan	Participant
Sawada, Nobutaka	Japan	Participant
Shigeyama, Masaru	Japan	Participant
Shimada, Kazumasa	Japan	Participant
Takada, Hiroko	Japan	Participant
Takagi, Tatsuhito	Japan	Participant
Takahashi, Hiroaki	Japan	Participant
Takeuchi, Jun	Japan	Participant
Teranishi, Koichi	Japan	Participant
Terasaki, Tomohiro	Japan	Participant
Torii, Honoka	Japan	Participant
Tanabe, Yuki	Japan	Participant
Tsubokura, Masaharu	Japan	Speaker

Watanabe, Tatsuki	Japan	Participant
Watanabe, Keiichi	Japan	Speaker
Yamada, Norikazu	Japan	Participant
Yamada, Tomoho	Japan	Participant
Yamamoto, Ayako	Japan	Participant
Yamana, Hajimu	Japan	Speaker
Yasuda, Yukiko	Japan	Participant
Yasuraoka, Satoru	Japan	Participant
Yoneyama, Shizuho	Japan	Participant
Yumoto, Keiichi	Japan	Speaker
Yamana, Hajimu	Japan	Participant
Yasuhiro, Baba	Japan	Participant
Bani Yasin, Abed Allah	Jordan	Participant
Hamdan, Khaleel Sami Mohammad	Jordan	Participant
Batyrbekov, Erlan	Kazakhstan	Participant
Vityuk, Vladimir	Kazakhstan	Participant
Onyango, Everlyne	Kenya	Participant
Bae, Youngmin	Korea, Republic of	Participant
Ha, Myeong Seon	Korea, Republic of	Participant
Jeong, Hun	Korea, Republic of	Participant
Kim, Jinjoo	Korea, Republic of	Participant
Kim, Kihyeon	Korea, Republic of	Participant
Kong, Byoung Moon	Korea, Republic of	Participant
Lee, Hyun Woo	Korea, Republic of	Participant
Lee, Sunmin	Korea, Republic of	Participant
Lee, Chul	Korea, Republic of	Participant
Lee, Geon Yong	Korea, Republic of	Participant
Lee, Hayeon	Korea, Republic of	Participant
Lee, Jin Ho	Korea, Republic of	Speaker
Park, Cheonkyeong	Korea, Republic of	Participant
Park, Seyoung	Korea, Republic of	Participant
Park, Eunyoung	Korea, Republic of	Participant
Tak, Minsoo	Korea, Republic of	Participant
Narbekov, Shailobek	Kyrgyzstan	Participant
Abbani, Bahlul	Libya	Participant
Abousuba Elturki, Ali	Libya	Participant
Elessawi, Khadra	Libya	Participant
Haj-Mohamed, Fathi Giuma Abdo	Libya	Participant
Malatim, Tariq	Libya	Participant
Ratemi, Wajdi Mohamed Sharef Ali	Libya	Participant
Shames, Husam Abdussalam Ramadan	Libya	Participant
Tantoush, Aisha	Libya	Participant
Ruzele, Paulius	Lithuania	Participant

Sabaitiene, Lina	Lithuania	Participant
Roes, Svenja	Luxembourg	Participant
Rabesiranana, Naivo	Madagascar	Participant
Bellamech, Nasser Dine	Mauritania	Participant
Mohamed Moussa, Ishagh	Mauritania	Participant
Hernandez Lopez, Hector	Mexico	Participant
Benider, Abdelkader	Morocco	Participant
Housni, Hafsa	Morocco	Participant
Marfak, Taib	Morocco	Participant
Mrabit, Khammar	Morocco	Speaker
Lwin, Khin Ye	Myanmar	Participant
Lwin, May Oo Khaing	Myanmar	Participant
Brugmans, Marco	Netherlands	Participant
Kuriene, Laimute	Netherlands	Participant
Philippens, Henry	Netherlands	Participant
Van Den Heuvel, Eline	Netherlands	Participant
Von Bolhuis, Annemiek	Netherlands	Participant
Lilly, David	New Zealand	Participant
Teo, Wanni	New Zealand	Participant
Adamu, John	Nigeria	Participant
Akor Solomon, Ekoja Okewu	Nigeria	Participant
Danbaba, Kenneth Luntsi	Nigeria	Participant
Muhammad, Zainab A.B.	Nigeria	Participant
Oyedokun, Olaide	Nigeria	Participant
Selemon, Deborah	Nigeria	Participant
Robinson, Carol	Norway	Participant
Solbakken, Even	Norway	Participant
Strand, Per	Norway	Participant
Al Brashdi, Khalid	Oman	Participant
Al Hinai, Ali	Oman	Participant
Arcilla, Carlo	Pakistan	Participant
Belmonte, Zachariah	Pakistan	Participant
Lazaro, Andrea Lynn	Pakistan	Participant
Maqbul, Naveed	Pakistan	Speaker
Refuerzo Lira, Pamela Eyre Victoria	Pakistan	Participant
Arcilla, Carlo	Philippines	Participant
Belmonte, Zachariah	Philippines	Participant
Lazaro, Andrea Lynn	Philippines	Participant
Refuerzo Lira, Pamela Eyre Victoria	Philippines	Participant
Glowacki, Andrzej	Poland	Participant
Mlynarkiewicz, Lukasz	Poland	Speaker
Portugal, Luis	Portugal	Participant
Al-Suwaidi, Hamda Sultan Sk	Qatar	Participant
Al-Maadeed, Ali	Qatar	Participant

Al-Sulaiti, Mohammed	Qatar	Participant
Al-Thani, Noof Ahmed	Qatar	Participant
Kamal, Rodah Fahad A I	Qatar	Participant
Barbos, Dumitru	Romania	Participant
Ciurea-Ercau, Cantemir Marian	Romania	Participant
Fako, Raluca Madalina	Romania	Participant
Holostencu, Sorin	Romania	Participant
Zalutchi, Silvian	Romania	Participant
Aparkin, Fedor	Russian Federation	Speaker
Bespala, Evgenii	Russian Federation	Participant
Bolgarov, Alexander	Russian Federation	Speaker
Chugunov, Vladimir	Russian Federation	Participant
Ferapontov, Alexey	Russian Federation	Speaker
Gorlova, Yulia	Russian Federation	Participant
Kozlov, Viacheslav	Russian Federation	Participant
Krasnov, Andrei	Russian Federation	Participant
Malinovskii, Pavel	Russian Federation	Participant
Meniailo, Ilia	Russian Federation	Participant
Metlyaev, Evgeny	Russian Federation	Participant
Milto, Ivan	Russian Federation	Participant
Petelin, Alexey	Russian Federation	Participant
Renev, Ivan	Russian Federation	Participant
Rakitskaya, Tatiana	Russian Federation	Participant
Romanov, Sergey	Russian Federation	Participant
Spoyalov, Oleg	Russian Federation	Participant
Titov, Vadim	Russian Federation	Speaker
Vasilev, Sergei	Russian Federation	Participant
Volgin, Alexandre	Russian Federation	Participant
Zhitov, Andrei	Russian Federation	Participant
Alsayyari, Fahad	Saudi Arabia	Participant
Al-Khomashi, Nasser	Saudi Arabia	Participant
Alrajhi, Abdulmajeed	Saudi Arabia	Participant
Velinov, Sladan	Serbia	Participant
A, Kannan	Singapore	Participant
Albert, Marc-Gérard	Singapore	Participant
Cheong, Denise	Singapore	Participant
Fan, Goh Si	Singapore	Participant
Hao, Tang Jia	Singapore	Participant
Har, Lau Wei	Singapore	Participant
Hui, Yeo Kiat	Singapore	Participant
Lam, Lui Cheuk	Singapore	Participant
Lee, Adrian	Singapore	Participant
Lim, Lewis	Singapore	Participant
Loh, Raymond	Singapore	Participant

Martina, Ira	Singapore	Participant
Miaosi, Chen	Singapore	Participant
Putra, Nur Azha	Singapore	Participant
Regalla, Manisha	Singapore	Participant
Ren, Than Yan	Singapore	Participant
S, Nivedita	Singapore	Participant
Seng, Lee Kheng	Singapore	Participant
Tan, Shou Qun	Singapore	Participant
Teo, Anabelle	Singapore	Participant
Tiong, Ng Boon	Singapore	Participant
Yap, Issac	Singapore	Participant
Zhong, Joanna	Singapore	Participant
Blazsekova, Marcela	Slovakia	Participant
Nano, Jan	Slovakia	Participant
Prazska, Milena	Slovakia	Participant
Rehak, Ivan	Slovakia	Participant
Subrtova, Natalia	Slovakia	Participant
Turner, Mikuláš	Slovakia	Participant
Ziakova, Marta	Slovakia	Speaker
Persic, Andreja	Slovenia	Participant
Skrk, Damijan	Slovenia	Participant
Zorko, Benjamin	Slovenia	Participant
Tyobeka, Bismark	South Africa	Speaker
Carrasco, Pilar	Spain	Participant
Carreter Ulecia, Mariano	Spain	Participant
Cuesta, Rodrigo	Spain	Participant
De Las Casas Fuentes, Alfonso	Spain	Participant
Guntirias Sanchez-Tori, Maria Luisa	Spain	Participant
Holguera Llamazares, Angel	Spain	Participant
Lentijo Lentijo, Juan Carlos	Spain	Participant
Lucio Carrasco, Pilar	Spain	Speaker
Monterrubio Villar, Esther	Spain	Participant
Martin, Rafael	Spain	Participant
Romera Gutierrez, Elvira	Spain	Speaker
Serena I Sender, Josep María	Spain	Participant
De Los Reyes Castelo, Alfredo	Spain	Speaker
Maheswarasarma, Doovaraha	Sri Lanka	Speaker
Raththagalage, Muditha Nayana Shantha Rathnayake	Sri Lanka	Participant
Danestig Sjögren, Catarina	Sweden	Speaker
Engkvist, Ingemar	Sweden	Participant
Cakir, Louisa Josefine	Switzerland	Participant
Frischknecht, Annina	Switzerland	Participant
Grote, Gudela	Switzerland	Speaker

Kenzelmann, Mark	Switzerland	Participant
Laggner, Benno	Switzerland	Participant
Mueller Germana, Annatina	Switzerland	Speaker
Nilsson, Hugo	Switzerland	Participant
Oberle, Markus	Switzerland	Participant
Rusch, Ronald	Switzerland	Participant
Sardella, Rosa	Switzerland	Speaker
Schwarz, Georg	Switzerland	Participant
Mirsaidzoda, Ilhom	Tajikistan	Participant
Sinsrok, Piyaporn	Thailand	Participant
Sriya, Maitree	Thailand	Participant
Pavenayotin, Niravun	Thailand	Participant
Promprasert, Natchapon	Thailand	Speaker
Sudprasert, Somjait	Thailand	Participant
Tumnoi, Yutthana	Thailand	Participant
Udomsomporn, Suchin	Thailand	Speaker
Songre, Douti Ardjoume	Togo	Participant
Akbas, Tahir	Turkey	Participant
Altintas, Burak	Turkey	Participant
Kilinç, Meltem	Turkey	Participant
Tkach, Taras	Ukraine	Participant
Gumenyuk, Dmytro	Ukraine	Participant
Novikov, Oleksandr	Ukraine	Speaker
Sapozhnykov, Iurii	Ukraine	Participant
Tokarskyi, Taras	Ukraine	Participant
Alharmoodi, Khaled	United Arab Emirates	Participant
Alkaabi, Hamad	United Arab Emirates	Participant
Almarzooqi, Hessa	United Arab Emirates	Speaker
Alnuaimi, Ahmed	United Arab Emirates	Participant
Alsaadi, Sara	United Arab Emirates	Participant
Kovacs, Emese	United Arab Emirates	Participant
Viktorsson, Christer	United Arab Emirates	Speaker
Zahradka, Dian	United Arab Emirates	Participant
Bibby, Liz	United Kingdom	Participant
Day, Katie	United Kingdom	Participant
Foy, Mark	United Kingdom	Speaker
Graham-Shaw, Kate	United Kingdom	Participant
Harvey, Penelope	United Kingdom	Speaker
Lawrie, Sarah	United Kingdom	Participant
Matthews, Benjamin	United Kingdom	Participant
Munchetty, Naga	United Kingdom	Moderator
Opperman, Bruce	United Kingdom	Participant
Reid, Michael	United Kingdom	Participant
Thomas, Gerry	United Kingdom	Speaker

Vaughan Jones, Hannah	United Kingdom	Moderator
Weightman, Michael	United Kingdom	Speaker
Weston, Rebecca	United Kingdom	Speaker
Matulanya, Machibya	United Republic of Tanzania	Participant
Crane-Roehrs, Melinda	United States of America	Participant
Cunningham, Kaylee Marie	United States of America	Participant
Disney, Patrick	United States of America	Participant
Dorman, Daniel	United States of America	Speaker
Franovich, Michael	United States of America	Speaker
Gunter, Linda	United States of America	Participant
Hanson, Christopher	United States of America	Speaker
Halpern, Clark	United States of America	Participant
Heinrich, Ann	United States of America	Participant
Jiang, Xiaodong	United States of America	Participant
Lacal, Maria	United States of America	Speaker
Mamish, Nader	United States of America	Participant
Marsh, Molly	United States of America	Participant
Micewski, Laura	United States of America	Participant
Nakanishi, Tony	United States of America	Participant
Perkins, Susan	United States of America	Speaker
Roberts, Darrell	United States of America	Speaker
Rogers, Hayden Page	United States of America	Moderator
Smith, Todd	United States of America	Speaker
Smith, Travis Scot	United States of America	Speaker
Wittick, Susan Van Lenten	United States of America	Participant
Mukhamedjanov, Aybek	Uzbekistan	Participant
Yusupov, Djalil	Uzbekistan	Participant
Abdukamilov, Shavkat	Uzbekistan	Participant
Shadmanov, Sardor	Uzbekistan	Participant
Al-Shami, Abdullah Ahmed Ali	Yemen	Participant
Esmail, Shadwan	Yemen	Participant
Mahdi, Mohammed	Yemen	Participant
Graham, Gerhard	CTBTO Preparatory Commission	Speaker
Garribba, Massimo	European Commission	Speaker
Noel, Marc	European Commission (EC), Joint Research Centre (JRC)	Speaker
Spasova, Yana	European Commission (EC), Joint Research Centre (JRC)	Participant
Blackburn, Carl Michael	Food And Agricultural Organization (FAO)	Speaker
Dercon, Gerd	Food And Agricultural Organization (FAO)	Speaker
Grossi, Rafael Mariano	International Atomic Energy Agency (IAEA)	Speaker

Evrard, Lydie	International Atomic Energy Agency (IAEA)	Speaker
Gaunt, Michael	International Labour Organization (ILO)	Participant
Pintado Nunes, Joaquim	International Labour Organization (ILO)	Speaker
Zodiates, Anastasios	International Labour Organization (ILO)	Participant
Garnier-Laplace, Jacqueline	OECD Nuclear Energy Agency (NEA)	Participant
Magwood, Iv, William	OECD Nuclear Energy Agency (NEA)	Speaker
Savel Ép Rouyer, Veronique	OECD Nuclear Energy Agency (NEA)	Participant
Muroya, Nobuhiro	OECD Nuclear Energy Agency (NEA)	Participant
Zimmermann, Moritz Richard	United Nations Scientific Committee on The Effects of Atomic Radiation (UNSCEAR)	Participant
Carr, Zhanat A.	World Health Organization (WHO)	Speaker
Neira, Maria	World Health Organization (WHO)	Speaker
Riishojgaard, Lars	World Meteorological Organization (WMO)	Speaker
Bilbao Y León, Sama	World Nuclear Association (WNA)	Speaker

ANNEX III. CONFERENCE SECRETARIAT AND PROGRAMME COMMITTEE

CHAIRPERSONS OF SESSIONS

Name	Role
M. Weightman, United Kingdom	President of the conference/Chairperson of the Session A
R. Velshi, Canada	Chairperson of the Session B
C. Hanson, United States of America	Chairperson of the Session C
N. Ban, Japan	Chairperson of the Session D
C.-M. Larsson, Australia	Chairperson of the Session E
M. Crane	Panel Discussion Moderator
N. Munchetty	Panel Discussion Moderator
H. Vaughan Jones	Panel Discussion Moderator

IAEA CONFERENCE SECRETARIAT

Name	Role
G. Caruso	Scientific Secretary
T. Danaher	Conference Coordination
A. Dixit	Outreach and Communication
E. Panteleymonova	
M. Kobein-Apolloner	Administrative Coordination and Support
P. Sales Barbosa	
K. Cottrell	
E. Lamce	
E. Freeman	Scientific Support
T. Karseka-Yanev	
M. Nikolaki	

INTERNATIONAL SCIENTIFIC PROGRAMME COMMITTEE

Name	Country
A. Gonzalez	Argentina
G. Hirth	Australia
R. Velshi	Canada
D. Drabova	Czech Republic

L. Evrard	France
J.-L. Lachaume	France
J. Regaldo	France
A.K. Dutta	India
S. Kaneko	Japan
F. Aparkin	Russian Federation
C. Viktorsson	United Arab Emirates
M. Weightman	United Kingdom
C. Hanson	United States of America
K. Svinicki	United States of America

LIST OF ABBREVIATIONS

ALPS	Advanced Liquid Processing System
CNS	Convention on Nuclear Safety
CTBTO	Preparatory Commission for The Comprehensived Nuclear-Test-Ban Treaty Organization
DOE	United States Department of Energy
EDF	Électricité de France
ENEC	Emirates Nuclear Energy Corporation
EPR	Emergency Preparedness and Response
EPReSC	Emergency Preparedness and Response Standards Committee
FANR	Federal Authority for Nuclear Regulation
FAO	Food and Agriculture Organization of the United Nations
GPS	Global Positioning System
IEC	Incident and Emergency Centre
IACRNE	Inter-Agency Committee on Radiological and Nuclear Emergencies
IAP	International Association for Public Participation
IDC	International Data Centre
ILO	International Labour Organization
IMS	International Monitoring System (for nuclear explosion monitoring)
JSGO	Japan Safeguards Office
LNT	Linear No-threshold
NEA	Nuclear Energy Agency
NGO	Non-governmental Organization
NPP	Nuclear Power Plant
NRA	Japan Nuclear Regulatory Authority
OECD	Organization for Economic Co-operation and Development
PNRA	Pakistan Nuclear Regulatory Authority
SMR	Small Modular Reactors
TEPCO	Tokyo Electric Power Company
UNSCEAR	United Nations Scientific Committee on the Effects of Atomic Radiation
WANO	World Association of Nuclear Operators
WGPC	Working Group on Public Communication of Nuclear Regulatory Organizations
WHO	World Health Organization
WMO	World Meteorological Organization

IAEA
International Atomic Energy Agency

ORDERING LOCALLY

IAEA priced publications may be purchased from the sources listed below or from major local booksellers.

Orders for unpriced publications should be made directly to the IAEA. The contact details are given at the end of this list.

NORTH AMERICA

Bernan / Rowman & Littlefield

15250 NBN Way, Blue Ridge Summit, PA 17214, USA
Telephone: +1 800 462 6420 • Fax: +1 800 338 4550

Email: orders@rowman.com • Web site: www.rowman.com/bernan

REST OF WORLD

Please contact your preferred local supplier, or our lead distributor:

Eurospan Group

Gray's Inn House
127 Clerkenwell Road
London EC1R 5DB
United Kingdom

Trade orders and enquiries:

Telephone: +44 (0)176 760 4972 • Fax: +44 (0)176 760 1640
Email: eurospan@turpin-distribution.com

Individual orders:

www.eurospanbookstore.com/iaea

For further information:

Telephone: +44 (0)207 240 0856 • Fax: +44 (0)207 379 0609
Email: info@eurospangroup.com • Web site: www.eurospangroup.com

Orders for both priced and unpriced publications may be addressed directly to:

Marketing and Sales Unit
International Atomic Energy Agency
Vienna International Centre, PO Box 100, 1400 Vienna, Austria
Telephone: +43 1 2600 22529 or 22530 • Fax: +43 1 26007 22529
Email: sales.publications@iaea.org • Web site: www.iaea.org/publications